Green Chemistry Series

Green Chemistry
Challenging Perspectives

Edited by
PIETRO TUNDO
epartment of Environmental Sciences, University of Ca' Foscari, Venice

and

PAUL ANASTAS
Industrial Chemistry Branch, US Environmental Protection Agency

Consorzio Interuniversitario Nazionale
"La Chimica per l'Ambiente"

Interuniversity Consortium
"Chemistry for the Environment"

OXFORD
UNIVERSITY PRESS

OXFORD
UNIVERSITY PRESS

Great Clarendon Street, Oxford OX2 6DP
Oxford University Press is a department of the University of Oxford.
It furthers the University's objective of excellence in research, scholarship,
and education by publishing worldwide in

Oxford New York

Athens Auckland Bangkok Bogota Buenos Aires Calcutta
Cape Town Chennai Dar es Salaam Delhi Florence Hong Kong Istanbul
Karachi Kuala Lumpur Madrid Melbourne Mexico City Mumbai
Nairobi Paris São Paulo Singapore Taipei Tokyo Toronto Warsaw
with associated companies in Berlin Ibadan

Oxford is a registered trade mark of Oxford University Press
in the UK and in certain other countries

Published in the United States
by Oxford University Press Inc., New York

© Oxford University Press, 2000
The moral rights of the author have been asserted
Database right Oxford University Press (maker)

First published 2000

All rights reserved. No part of this publication may be reproduced,
stored in a retrieval system, or transmitted, in any form or by any means,
without the prior permission in writing of Oxford University Press.
or as expressly permitted by law, or under terms agreed with the appropriate
reprographics rights organization. Enquiries concerning reproduction
outside the scope of the above should be sent to the Rights Department.
Oxford University Press, at the address above

You must not circulate this book in any other binding or cover
and you must impose this same condition on any acquirer

A catalogue record for this book is available from the British Library
Library of Congress Cataloging in Publication Data

Green chemistry : challenging perspectives / edited by Pietro Tundo and Paul Anastas.
1. Environmental chemistry–Industrial applications. 2. Environmental management.
I. Tundo, Pietro, 1945- II. Anastas, Paul T., 1962- III. Series.
TP155.2.E58 G74 2000 660–dc21 99-052845

ISBN 0 19 850455 1

Typeset by EXPO Holdings, Malaysia
Printed in Great Britain
on acid-free paper by
T.J. International Ltd, Padstow, Cornwall

Preface

In 1996, the Working Party on 'Synthetic Pathways and Processes in Green Chemistry' was established within the IUPAC Commission III.2 to bring attention to an area of rapidly expanding research and development, defined as *'the invention, design and application of chemical products and processes to reduce or to eliminate the use and generation of hazardous substances'*.

About a year later, from September 28 to October 1 1997, chemists from 12 countries assembled in Venice for the first Green Chemistry Confrence on the European Continent, at the invitation of the Italy's Interuniversity Consortium 'Chemistry for the Environment'.

The 'Green Chemistry: Challenging Perspectives' Conference was the first European opportunity for chemists from different fields (industrial, academic, governmental and non governmental organizations) dealing with environmentally benign chemical synthesis and processing, to present their research and to establish collaborations aimed at the adoption of the green chemistry approach in industrial practice. A number of contributions selected from the board array of presentations featured by the Venice Conference are reported in this book.

The design and development of new cleaner alternative synthetic pathways, such as catalysis (Chapters 9 and 10) and bio-catalysis (Chapters 2, 6, and 13), as well as of new cleaner processes, such as photochemistry (Chapters 1 and 7) and biomimethic syntheses (Chapter 4) are comprehensively addressed.

New optimized reaction conditions, and control systems and catalysts, for the existing syntheses and processing of chemicals are treated in Chapter 5; alternative less toxic solvents are reported in Chapter 8, while Chapters 3 and 8 tackle solvent-free reactions.

Finally, an example of new chemicals with novel structures, that reduce or eliminate toxicity and hazard while maintaining efficacy, is reported in Chapter 11.

This book will serve not only to extend the perspective of current practitioners of green chemistry to a more global context, but also to transfer the concepts of environmentally benign chemical synthesis to scientists who are currently unaware of its existence or potential relevance. The development of new processes that are simultaneously economically sustainable and environmentally responsible is the challenge for the twenty-first century.

Venice P.T.
November 1999

Contents

List of contributors ix

1. **M. Anpo**
   Application of titanium oxide photocatalysts to
   improve our environment   1

2. **E. Bolzacchini, S. Meinardi, M. Orlandi,
   B. Rindone, G. Brunow, P. Pietikainen,
   and P. Rummakko**
   'Green oxidations': horseradish peroxidase
   (HRP)-catalyzed regio and diastereoselective
   preparation of dilignols   21

3. **F. Bigi, R. Maggi, and G. Sartori**
   Fine chemicals preparation via solventless reactions
   under heterogeneous catalysis   37

4. **G. Carturan, R. Campostrini, and R. Dal Monte**
   Secondary metabolites from cells immobilized by
   a SiO_2 sol–gel layer   61

5. **T.J. Collins, J. Hall, L. Vuocolo, N.L. Fattaleh,
   I. Suckling, C.P. Horwitz, S.W. Gordon-Wylie,
   R.W. Allison, T.J. Fullerton, and L.J. Wright**
   The activation of hydrogen peroxide for selective,
   efficient wood pulp bleaching   79

6. **E. D'Addario, C. Colapicchioni, E. Fascetti,
   R. Gianna, A. Robertiello**
   Microbial desulfurization of petroleum derivatives   107

7. **A. Maldotti, R. Amadelli, A. Molinari, V. Carassiti**
   Cyclohexane oxygenation with inorganic photocatalysts 125

8. **F. Montanari, G. Pozzi, S. Quici**
   Catalytic hydrocarbons oxidation under fluorous/organic two-phase conditions 145

9. **A. Corma and J.M. López Nieto**
   Strong acid solid materials as alternative catalysts in isobutane/2-butene alkylation 163

10. **L. Prati and M. Rossi**
    Selective catalytic oxidation of 1,2-diols in alkaline solution: an environmentally friendly alternative for α-hydroxy-acids production 183

11. **F. Rivetti**
    Dimethylcarbonate: an answer to the need for safe chemicals 201

12. **R.S. Varma**
    Environmentally benign organic transformations using microwave irradiation under solvent-free conditions 221

13. **A. Schmid, B. Witholt**
    Organic chemistry via biocatalysis 245

    Index 267

Contributors

R.W. Allison; PAPRO New Zealand Limited, Private Bag 3020, Rotorua, New Zealand

R. Amadelli; Dipartimento di Chimica e Centro di Studio su Fotoreattività e Catalisi del CNR, Università degli Studi di Ferrara, Via L. Borsari 46, 44100 Ferrara, Italy

M. Anpo; Department of Applied Chemistry, College of Engineering, Osaka Prefecture University, 1-1 Gakuen-cho, Sakai, Osaka 599–8531, Japan

F. Bigi; Dipartimento di Chimica e industriale dell'Università, Viale delle Scienze, 1-43100 Parma, Italy

E. Bolzacchini; Dipartimento di Scienze dell'Ambiente e del Territorio, Università di Milano, Via Emanueli 15, I-20126 Milano, Italy

G. Brunow; Department of Chemistry, University of Helsinki, P.O. Box 55, Fin-00014 Helsinki, Finland

R. Campostrini; Department of Materials Engineering, University of Trento, 38050 Mesiano, Italy

V. Carassiti; Dipartimento di Chimica e Centro di Studio su Fotoreattività e Catalisi del CNR, Università degli Studi di Ferrara, Via L. Borsari 46, 44100 Ferrara, Italy

G. Carturan; Department of Materials Engineering, University of Trento, 38050 Mesiano, Italy

C. Colapicchioni; Eniricerche, 00016 Monterotondo (Rome), Italy

T.J. Collins; Department of Chemistry, Carnegie Mellon University, 4400 Fifth Avenue, Pittsburgh, PA 15213, USA

A. Corma; Instituto de Tecnologia Quìmica, UVP-CSIC, Avda, Los Naranjos s/n, 46022-Valencia, Spain

E. D'Addario; Eniricerche, 00016 Monterotondo (Rome), Italy

R. Dal Monte; R & C Scientifica, 36077 Altavilla Vicentina, Italy

E. Fascetti; Eniricerche, 00016 Monterotondo (Rome), Italy

N.L. Fattaleh; Department of Chemistry, Carnegie Mellon University, 4400 Fifth Avenue, Pittsburgh, PA 15213, USA

T.J. Fullerton; PAPRO New Zealand Limited, Private Bag 3020, Rotorua, New Zealand

R. Gianna; Eniricerche, 00016 Monterotondo (Rome), Italy

S.W. Gordon-Wylie Department of Chemistry, Carnegie Mellon University, 4400 Fifth Avenue, Pittsburgh, PA 15213, USA

J. Hall; Department of Chemistry, University of Auckland, Private Bag, Auckland 1, New Zealand

C.P. Horwitz; Department of Chemistry, Carnegie Mellon University, 4400 Fifth Avenue, Pittsburgh, PA 15213, USA

J.M. López Nieto; Instituto de Tecnologia Quimica, UVP-CSIC, Avda, Los Naranjos s/n, 46022-Valencia, Spain

R. Maggi; Dipartimento di Chimica e Industriale dell'Università, Viale delle Scienze, 1-43100 Parma, Italy

A. Maldotti; Dipartimento di Chimica e Centro di Studio su Fotoreattività e Catalisi del CNR, Università degli Studi di Ferrara, Via L. Borsari 46, 44100 Ferrara, Italy

S. Meinardi; Dipartimento di Scienze dell'Ambiente e del Territorio, Universita di Milano, Via Emanueli 15, I-20126 Milano, Italy

A. Molinari; Dipartimento di Chimica e Centro di Studio su Fotoreattività e Catalisi del CNR, Università degli Studi di Ferrara, Via L. Borsari 46, 44100 Ferrara, Italy

F. Montanari; Dipartimento di Chimica Industriale dell'Universita e Centro CNR di Studio sulla Sintesi e Stereochimica di Speciali Sistemi Organici, Via Golgi 19, I-20133 Milano, Italy

M. Orlandi; Dipartimento di Scienze dell'Ambiente e del Territorio, Universita di Milano, Via Emanueli 15, I-20126 Milano, Italy

P. Pietikainen; Department of Chemistry, University of Helsinki, P.O. Box 55, Fin-00014 Helsinki, Finland

G. Pozzi; Dipartimento di Chimica Industriale dell'Universita e Centro CNR di Studio sulla Sintesi e Stereochimica di Speciali Sistemi Organici, Via Golgi 19, I-20133 Milano, Italy

L. Prati; Dipartimento di Chimica Inorganica Metallorganica e Analitica e Centro CNR, via Venezian 21, 20133 Milano, Italy

S. Quici; Dipartimento di Chimica Industriale dell'Università e Centro CNR di Studio sulla Sintesi e Stereochimica di Speciali Sistemi Organici, Via Golgi 19, I-20133 Milano, Italy

B. Ridone; Dipartimento di Scienze dell'Ambiente e del Territorio, Universita di Milano, Via Emanueli 15, I-20126 Milano, Italy

F. Rivetti; EniChem, Divisione Intermedi, Via Maritano 26, 20097 S. Donato Milanese MI, Italy

A. Robertiello; Eniricerche, 00016 Monterotondo (Rome), Italy

M. Rossi; Dipartimento di Chimica Inorganica Metallorganica e Analitica e Centro CNR, via Venezian 21, 20133 Milano, Italy

P. Rummakko; Department of Chemistry, University of Helsinki, P.O. Box 55, Fin-00014 Helsinki, Finland

G. Sartori; Dipartimento di Chimica e Industriale dell'Università, Viale delle Scienze, 1-43100 Parma, Italy

A. Schmid; Institut für Biotechnologie, ETH Hönggerberg, 8093 Zürich, Switzerland

I. Suckling; PAPRO New Zealand Limited, Private Bag 3020, Rotorua, New Zealand

R.S. Varma; U.S. Environmental Protection Agency, National Risk Management Research Laboratory, 26 West Martin Luther King Drive, MS 443, Cincinnatti, Ohio 45268, USA

L.D. Vuocolo; Department of Chemistry, Carnegie Mellon University, 4400 Fifth Avenue, Pittsburgh, PA 15213, USA

B. Witholt; Institut für Biotechnologie, ETH Hönggerberg, 8093 Zürich, Switzerland

L.J. Wright; Department of Chemistry, University of Auckland, Private Bag, Auckland 1, New Zealand

1 Application of titanium oxide photocatalysts to improve our environment

Masakazu Anpo

1.1 Introduction

Environmental pollution and destruction on a global scale have drawn attention to the vital need for new environmentally friendly, clean chemical technologies and processes. The development of these areas is the most important challenge facing chemical scientists in green chemistry. Strong contenders as environmentally harmonious or friendly catalysts are solid photocatalysts which will be able to work at room temperature and in a clean and safe manner.[1]

Since the oil shock of the early 1970s, many studies have been carried out on the reactivity of solid photocatalysts such as titanium dioxide semiconductor powders with the aim of converting light energy, most ideally solar energy, into chemical energy.[2] However, to attain such a goal using titanium dioxide (TiO_2) semiconductors, strong and highly powerful light sources emitting large numbers of ultraviolet photons have been required. Therefore, although it is already possible to decompose water into H_2 and O_2 using semiconducting photocatalysts such as Pt loaded titanium dioxide (Pt/TiO_2), it is still difficult to industrialize these processes to produce H_2 from H_2O from an economic point of view. Some studies have been carried out in this field to improve the efficiency of the reactions. In fact, we have carried out studies on the development of TiO_2 photocatalysts which are able to decompose and purify various pollutants such as NOx in the atmosphere and toxic compounds dissoloved in water even under weak light irradiation.[3-6]

Although further fundamental research will be required for the effective and widespread application of photocatalysts, we can already see practical applications in use on a small scale.[7-8] Notably, active research using a variety of photocatalysts can be seen in areas concerning the decontamination of the toxic compounds in polluted water, the photodetoxification of atmosphere toxic compounds as well as the decomposition of compounds causing offensive odors in the atmosphere, the direct decomposition of NOx into nitrogen and oxygen and the reduction of carbon dioxide with water using photocatalysts for the effective utilization of limited carbon resources.[6,9,10]

In this chapter, current developments in photocatalytic research on the TiO_2 catalyst, which is considered the most thermodynamically stable and photocatalytically reactive, have been discussed, specifically for environmental concerns.

1.2 Fundamentals of photocatalysis

As shown in Fig. 1.1a and 1.1b, when semiconducting powdered TiO_2 catalysts are irradiated with light that is greater than their bandgap energy of about 3.2 eV for anatase type TiO_2, electrons in the valence band are excited and electrons are injected into the conduction band while holes are produced in the valence band. Thus, produced electrons have a highly reactive reduction potential while the holes have a highly reactive oxidation potential which together induce a catalytic reaction on the surface of the TiO_2 catalysts. Such reactions are referred to as photocatalytic reactions and the catalysts which are able to induce such surface catalytic reactions are referred to as in photocatalysts.[2,3,6,7]

As Fig. 1.1c shows for the photosynthesis processes in green plants, photosynthetic reactions can also be understood in terms of reactions caused by electrons and holes which are generated by light energy. In this respect, photocatalysis may also be referred to as artificial in photosynthesis. It differs dramatically from the traditionally utilized thermal catalysing reactions (induced by the application of environmentally hazardous heat energy generated by burning fossil fuels) in that photocatalytic reactions proceed under moderate conditions at room

Application of titanium oxide photocatalysts to improve our environment

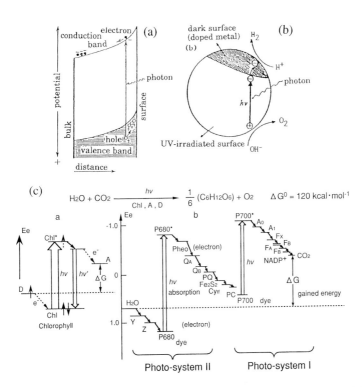

Fig. 1.1 Schematic representation of the photoexcitation of semiconducting TiO_2 powdered catalysts (a), photo-formation of electrons (e^-) and holes (h^+), their charge separation, and the reduction of H^+ and oxidation of OH^- to produce H_2 from H and O_2 via the formation of OH radicals in an aqueous system (b), and photosynthesis mechanism involving photosystem I and II (c).

temperature, normal atmospheric pressure and light irradiation, making photocatalysis an efficient, clean and safe chemical process.[1,2,7]

1.3 The application of photocatalysis for environmental detoxification

Environmental problems for which catalytic technology is urgently being developed include the decontamination of polluted water, the purification of exhaust gases from various industries and transportation vehicles, the decomposition of offensive atmospheric odors as well as toxins, the fixation of carbon dioxide and the decomposition of NOx, SOx and chlorofluorocarbons. In particular, as mentioned above, photocatalytic reactions can be usefully applied to treat very low and dilute concentrations of toxic reactants in water and the atmosphere on a large scale.

The practical, widespread use of TiO_2 photocatalysts can be seen in various applications and materials, in which TiO_2 thin films or micropowdered photocatalysts are fixed onto supports such as active carbon, transparent glass, zeolites, cement, tile, textile fabrics, and even onto paper. These supports have adsorption properties more or less attributed to their large surface areas, which enable the photocatalytic reaction systems to condense the concentration of the very dilute reactants achieving highly efficient reactions. In fact, recent applications of TiO_2 thin films show that photocatalytic reactions for the decontamination of polluted air proceed with sufficient efficiency even with the use of weak room light.[8]

1.3.1 The atmosphere

1.3.1.1 Elimination of nitrogen oxides (NOx such as NO and N_2O)

To create a comfortable living environment, it is crucial to remove NOx and SOx in the atmosphere, especially from highways, tunnels and highly populated areas, using chemical processes which are not dangerous, are efficient and clean and work without any loss of energy. The ion-exchanged copper/ZSM-5 zeolite catalyst has attracted a great deal of attention as a potential catalyst for the direct decomposition of NOx into N_2 and O_2. This catalyst can decompose NOx into N_2 and O_2 in the temperature ranges of 673–773 K. On the other hand, as shown in Fig. 1.2, UV irradiation of the copper(I)/ZSM-5 zeolite catalyst in the presence of NO leads to the formation of N_2 and O_2 with a good linear relationship between the irradiation time and the conversion of NO, even at temperatures as low as 275 K.[4,11,12] Thus, under UV irradiation ($\lambda > 280$ nm), the Cu(I)/ZSM-5 zeolite works as a photocatalyst to decompose NO into N_2 and O_2 with a good stoichiometry ($N_2: O_2 = 1:1$). However, in the copresence of O_2 both the catalytic and photocatalytic reactivities of the copper/ZSM-5 zeolite catalyst are dramatically suppressed.

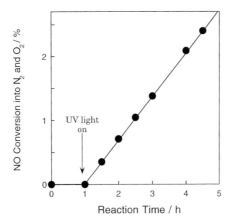

Fig. 1.2 Photocatalytic decomposition of NO into N_2 and O_2 on the Cu^+/ZSM-5 zeolite catalyst under UV irradiation of the catalyst in the presence of NO at 275 K.

However, fortunately, it has been observed that UV-irradiation ($\lambda > 220$ nm) of the Ag(I), with its electronic state being similar to that of Cu(I)ion, anchored onto the ZSM-5 zeolite (Ag(I)/ZSM-5 catalyst) in the presence of NO leads, to the direct photocatalytic decomposition of NO into N_2, O_2 and N_2O at 275 K. The photocatalytic activity of the Ag(I)/ZSM-5 catalyst is 20 times higher than that of the Cu(I)/ZSM-5 catalyst, which can be attributed to the much higher energy level of the excited state of the Ag(I)ion than that of the Cu(I)ion. Of special interest is that the Ag(I)/ZSM-5 zeolite catalyst operates as a photocatalyst even in the copresence of O_2 and/or H_2O.[11,12]

Furthermore, UV irradiation ($\lambda > 280$ nm) of the highly dispersed titanium oxides anchored onto zeolites by an ion-exchanged method or incorporated into zeolite frameworks in the presence of NO, have been found to produce N_2 and O_2 with a selectivity much higher than 80%, showing a sharp contrast to the photocatalytic decomposition reaction of NO on TiO_2 semiconducting powdered catalysts, where more than 80% of the product is N_2O and the selectivity for the direct decomposition NO into N_2 is less than 20%.[13]

A detailed analysis of the local structures of the catalysts using various molecular spectroscopies, such as XAFS (XANES and FT-EXAFS) and photoluminescence, has revealed that there is an interesting

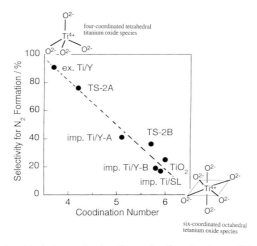

Fig. 1.3 Relationship between the local coordination structures of the titanium oxide photocatalysts and the selectivity for the formation of N_2 and O_2 in the photocatalytic decomposition of NO on the various titanium oxide photocatalysts under UV irradiation of the catalysts in the presence of NO at 295 K.

relationship between the local structures of the titanium oxide photocatalysts and the efficiency (as well as the selectivity) for the decomposition of NO into N_2 and O_2. As shown in Fig. 1.3, the titanium oxide catalysts having tetrahedral co-ordination of a TiO_4 unit structure exhibit a high photocatalytic reactivity and a high selectivity for the direct decomposition of NO into N_2 and O_2. Of special interest is that such highly dispersed titanium oxide photocatalysts having a tetrahedral co-ordination operate as an efficient photocatalyst even in the copresence of O_2.[4,5,13]

These results clearly indicate that these Ag(I)/ZSM-5 zeolite catalysts and highly dispersed titanium oxide catalysts anchored onto zeolite or incorporated into zeolite framework are one of the promising candidates for utilization as an environment-saving photocatalyst for the removal of NOx in the atmosphere.

On the other hand, the utilization of the TiO_2 semiconducting powdered catalysts supported on active carbon has been proposed by Ibusuki and Takeuchi for the removal of NOx even in the presence of O_2 and gaseous H_2O.[14] Such utilization of the TiO_2 is expanded onto various support

Application of titanium oxide photocatalysts to improve our environment

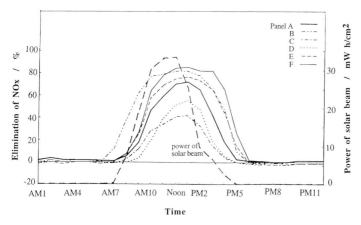

Fig. 1.4 The reaction time profiles of the elimination of NO in the atmosphere in a day using the TiO$_2$ photocatalytic system.

materials such as cement and SiO$_2$, which have a large surface area, and these TiO$_2$ photocatalytic systems are now being applied in Japan as one of the most promising ways for the removal of NOx near the most heavily travelled highways. In these photocatalytic oxidation reaction systems, NO is photocatalytically oxidized into NO$_2$, being trapped on the support moeity, and then NO$_2$ is again photocatalytically oxidized into NO$_3^-$, and then finally this NO$_3^-$ species is altered into HNO$_3$ by H$_2$O. Thus, ultimately NO can be altered into HNO$_3$ by rain.

According to the results obtained by the fieldwork tests using various types of photocatalytic TiO$_2$ panels which were placed under a main expressway in a heavily congested road area of Osaka, as can be seen in Fig. 1.4, when the sun rises, the elimination of NOx starts and increases with the time, passing through a maximum at around 2 PM, then decreases and finally the elimination reaction completely ceases when the sun sets. Photocatalytically formed NO$_3^-$ is removed as HNO$_3$ from the TiO$_2$ panels when it is raining.[15] In many big cities in Japan, such fieldwork using various types of photocatalytic TiO$_2$ panels has been carried out and this system may actually be applied for the elimination of NOx, especially in areas of high traffic surrounded by tall buildings.

1.3.1.2 Removal of offensive odors and disinfection

As has been mentioned earlier, depending on the reaction, in some cases, a weak ultra-violet room light may be sufficient to produce the desired results. In fact, it is possible to decompose compounds causing offensive odors and bacteria through photocatalytic oxidation reactions and to disinfect using TiO_2 catalysts prepared on walls and window panes which then act as a photocatalyst even with the weak room light as the UV irradiation light source. Using this method, building materials will be able to incorporated the photocatalytic functions of antifouling, antibacterial and odor removal under normal conditions in everyday living areas in which only a very weak room light is available as the light source.[8]

1.3.1.3 Reduction of carbon dioxide with water

The reduction of carbon dioxide with water is a so-called energy storing 'up-hill' reaction in which the progress of the reaction is, thermodynamically, extremely difficult. However, in actual experiments using TiO_2 photocatalysts, the production of HCOOH, HCHO, CH_3OH and CH_4 as a result of the reduction of carbon dioxide with water using various types of photocatalysts is possible and has been observed and reported by a number of groups.[4,5,6,16,17]

On the semiconducting powdered TiO_2 photocatalysts, only a small amount of CH_4 formation has been observed in the reduction reaction of CO_2 with gaseous H_2O at room temperature. On the other hand, on highly dispersed titanium oxide catalysts anchored onto porous glass or onto zeolites by CVD or an ion-exchange method as well as on titanium silicalite catalysts such as TS-1, Ti-MCM-41 and Ti-MCM-48, the formation of CO, CH_4, CH_3OH and other hydrocarbons can be observed in the photocatalytic reaction of CO_2 with gaseous H_2O.[5,6,18,19] As shown in Fig. 1.5, the photocatalytic reduction reaction of CO_2 with H_2O proceeds with a relatively high efficiency having a good linearity against the UV-irradiation time in the temperature ranges of 298–323 K. Furthermore, in the photocatalytic reduction reaction of CO_2 with gaseous H_2O on the Ti/Si binary oxide catalysts prepared by the sol-gel method from a mixture of $Ti(OBu)_4$ and $Si(OEt)_4$, a high yield and selectivity in the CH_3OH formation can be achieved at 323 K.

The *in situ* photoluminescence, ESR, IR and XAFS (XANES and FT-EXAFS) measurements of the catalysts clearly show that only catalysts having Ti(IV) ions in tetrahedral co-ordination exhibit photolumines-

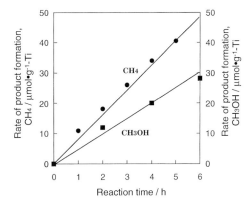

Fig. 1.5 The reaction time profiles for the formation of CH_4 and CH_3OH in the photocatalytic reduction of CO_2 with H_2O on the Ti-MCM-48 mesoporous zeolite at 323 K.

cence in the absence of reactant molecules as a radiative decay process from the charge transfer excited state of titanium oxide, $(Ti^{3+}-O^{-})^{*}$ and also exhibit photocatalytic reactivity for the reduction of CO_2 with H_2O to produce CH_4, CH_3OH and CO. Thus, the photocatalytic reduction of CO_2 with H_2O is closely linked to the much higher reactivity of the charge transfer excited state of the titanium oxide, $(Ti^{3+}-O^{-})^{*}$ due to the presence of a well-dispersed, homogeneous titanium oxide species on the surface.[8,12] Furthermore, the total yield of the photo-formed CH_4, CH_3OH and CO was found to increase when UV irradiation was carried out at 323 K as compared with irradiation at 275 K, suggesting the joint effect of UV light and heat in the photocatalytic reactions. These results on the photocatalytic reduction of CO_2 with H_2O provide important and valuable information for the design of highly efficient and active photocatalytic systems.[3–6,16]

1.3.2 Aqueous solutions

1.3.2.1 Hydrogen production using photocatalysts

The evolution of H_2 from water is one of the most important catalytic reactions. It takes place in the following oxidizing and reduction reactions:

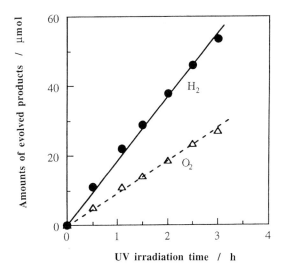

Fig. 1.6 The complete and stoichiometric photocatalytic decomposition of liquid H_2O into H_2 and O_2 using a Ti-B binary oxide photocatalyst under UV irradiation at 295 K.

$$\text{Oxidizing reaction:} \quad 2H_2O + 4h^+ \rightarrow 4H^+ + O_2$$
$$\text{Reduction reaction:} \quad 4H^+ + 4e^- \rightarrow 2H_2$$

The production of H_2 from water has been reported using a number of semiconducting photocatalysts such as TiO_2 and $SrTiO_3$.[20] However, the reaction is not efficient and the quantum yield is still in the low range of 10^{-2} to 10^{-3}. The main reason for the low efficiency of water splitting is that the reverse reaction of H_2 and O_2 takes place easily on the catalyst. However, as reported by Domen *et al.*,[21] if the layered $K_4Nb_6O_{17}$ catalyst having small amounts of NiO (NiO/$K_4Nb_6O_{17}$ (bandgap value = 3.3 eV) is applied as the photocatalyst, the water can be completely decomposed into H_2 and O_2 at an efficiency of approximately 12%. Because the mechanism is such that H_2 is produced form the NiO moiety between the layers and O_2 is produced form the catalytic layers of the $K_4Nb_6O_{17}$, with this layered photocatalyst, the reverse reaction is suppressed and H_2 and O_2 is formed very efficiently. Recently, as shown in Fig. 1.6, Moon *et al.* have found that titanium oxides partly covered by boron oxide moeity decompose water into H_2 and O_2 in a relatively high efficiency under UV irradiation in aqueous solution at 295 K.[22] They have found that in these photocatalysts, boron-oxides moeity suppress the reverse reaction of $H_2 + O_2$ into H_2O, resulting in a high efficiency for the production of H_2 and O_2 from water.

It is known that the production of H_2 from H_2O using sacrificial agents such as alcohols is extremely enhanced by doping small amounts of Pt or Pd into TiO_2 photocatalysts. The H_2 production yields from aqueous solutions of CH_3OH or C_2H_5OH improves dramatically by the addition of small amounts of Pt or Pt and RuO_2 onto TiO_2 photocatalyst.[2,6,7] Such highly efficient H_2 production in these reaction systems through photocatalysis can be applied to many other organic compounds dissolved in water. However, these photocatalytic reaction systems are still difficult to industrialized and require an improvement in quantum yield, at least up to around 0.1–0.3.

1.3.2.2 Decontamination of polluted water

Utilizing the high reduction ability and high oxidizing ability of the photoformed electrons and holes in the photocatalysts, various harmful substances dissolved in water can be disintegrated and decomposed into harmless compounds. Table 1.1 shows the time taken to decompose the many harmful substances dissolved in water into half the original concentration using a TiO_2 photocatalyst.[7] The disintegration time differs according to the compound. However, it can be seen that even extremely stable yet harmful substances such as dioxin can be decomposed. The time taken for the harmful substances to decompose can be greatly improved by supplying oxygen into the photocatalytic

Table 1.1 UV irradiation times of the TiO_2 photocatalyst to decompose the reactants to half the original concentrations.

Compounds	Concentration (ppm)	Irradiation time (min)
4-chlorophenol	6.0	14
2,4,5-trichlorophenol	20.0	55
pentachlorophenol	12.0	20
1,2,4-trichlorobenzene	10.0	24
3,3'-dichlorobiphenyl	1.0	10
2,7-dichlorodibenzo-p-dioxin	0.2	46

reaction systems or by adding hydrogen peroxide. The photocatalytic reaction in the solution involves the oxidation processes by highly active OH radicals which are produced by the oxidation of OH^- by the holes formed by light irradiation. The electrons then reduced the oxygen to produce O_2^-. The high reactivity of these OH radicals and O_2^- allow the highly efficient oxidative decomposition of harmful substances in solution.

The degradation of impurities, including harmful substances such as PCBs and dioxin in water, the decomposition of harmful substances such as detergents and fertilizers and the treatment of substances such as industrial waste liquids using photocatalysts will be the first areas in which photocatalysis will be practically applied in a large scale. Furthermore, the manufacture of semiconducting devices requires extremely pure water and such photocatalytic reaction systems may be effective since such water purification is difficult with other methods. Currently, in these areas, the separation of the photocatalysts from water after use, its recycling and maintaining its effectiveness or lifetime are being reviewed as key issues facing practical applications of photocatalysts on a large scale.

Liquid phase photocatalytic reactions by applying anchoring or grafting, or fixing titanium oxide photocatalysts onto thin films, oxide compounds, on the surface of a transparent sheet such as Vycor glass or optical fibers using the sol-gel or CVD method is also being widely investigated.[4,6] The objectives of research involving environment-enhancing photocatalysis are the separation of the catalysts from the system after use as well as the large-scale treatment of dilute concentrations of reactants in water and the atmosphere. With regard to such objectives, the possibility of fixing a photocatalyst onto substances such as active carbon with high adsorption properties attributed to its large surface area is actively being investigated in order to condense the concentration of the reactants, making them highly reactive.

1.4 Utilization of solar beams or visible light

As has been discussed, there are no limits to the possibilities and applications of TiO_2 photocatalysts and photocatalytic reaction systems as an 'environmentally harmonious catalyst'. However, as can be seen in Fig. 1.7 and Table 1.2 and unlike photosynthesis in green plants, the

Application of titanium oxide photocatalysts to improve our environment 13

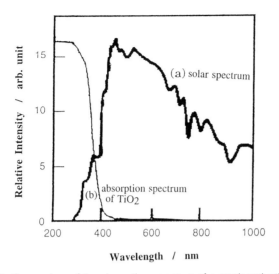

Fig. 1.7 Comparison of the absorption spectrum of a semiconducting TiO_2 powdered photocatalyst and the solar spectrum on the surface of the earth.

TiO_2 photocatalyst in itself does not allow the use of visible light and can only make use of 3–5% of the solar beams that reach the earth. Therefore, to establish clean and safe photocatalytic reaction systems using the most chemically and environmentally ideal energy source,

Table 1.2 Various semiconducting materials, their bandgap energies and the corresponding wavelengths as well as thermal energies.

Semiconductors	bandgap (eV)	Wavelength (nm)	Energy (kcal mol^{-1})
SnO_2	3.8	326	87.7
ZnS	3.6	346	83.1
ZnO	3.2	388	73.8
WO_3	3.2	388	73.8
TiO_2	3.2	388	73.8
$SrTiO_3$	3.2	388	73.8
SiC	3.0	413	69.2
CdS	2.5	496	57.7
Fe_2O_3	2.3	539	53.1
Gap	2.25	551	51.9
CdSe	1.7	730	39.2

14 *Green Chemistry: Challenging Perspectives*

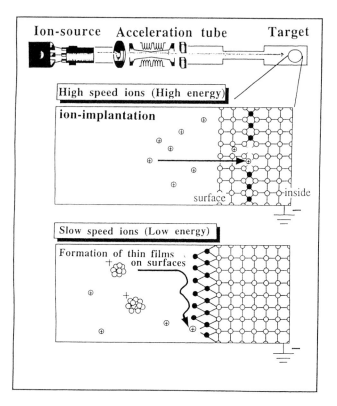

Fig. 1.8 Schematic illustration of principles of the advanced metal ion-implantation method.

solar light, it is vital to develop TiO_2 photocatalysts which can operate with high efficiency under visible light irradiation.[6,23–26]

We have applied the metal ion-implantation method to modify the electronic properties of semiconductors by bombarding them with high energy metal ions and we have found that this advanced physical method is the most suitable and promising for the dramatic modification of the electronic state of the TiO_2 photocatalysts. Figure 1.8 shows an illustration of the advanced metal ion-implantation method.

As shown in Fig. 1.9, the absorption band of the TiO_2 photocatalyst was found to shift smoothly to visible light regions, the extent of the red

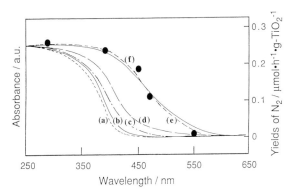

Fig. 1.9 UV–Vis absorption spectra (diffuse reflectance) of the unimplanted original TiO_2 photocatalyst (a) and the Cr ion-implanted TiO_2 photocatalysts (b–e), and the action spectrum of the photocatalytic reaction on the Cr ion-implanted TiO_2 (f) corresponding to the TiO_2 photocatalyst of spectrum (e). The amounts of Cr ions-implanted (in 10^{-7} mol/g): b, 2.2; c, 6.6; d, 13; e, 26.

shift depending on the amount and type of metal ions implanted (spectra b–e), with the absorbance maximum and minimum values always remaining constant. The TiO_2 photocatalysts implanted with Cr or V ions were found to exhibit the largest shift to visible light regions up to a wavelength of 550–600 nm. However, the implantation of Mg ions, Ti ions or Ar ions scarcely changed the absorption band of TiO_2. We have found that metal ions implanted within the bulk of the catalyst modified the electronic properties of TiO_2 in which the d-electrons of the metal ions implanted may be associated with this phenomenon, showing that photocatalytic reactions take place under visible light.[23–27]

As mentioned above, normally with TiO_2 photocatalysts, the photocatalytic reaction does not proceed under visible light irradiation ($\lambda > 450$ nm). However, we have found that visible light irradiation ($\lambda > 450$ nm) of these metal ion-implanted TiO_2 photocatalysts led to various significant photocatalytic reactions. As shown in Fig. 1.10, visible light irradiation of the Cr ion-implanted TiO_2 photocatalyst in the presence of NO at 275 K led to the decomposition of NO into N_2O as well as N_2 and O_2, with a good linearity against the irradiation time. Under the same conditions of visible light irradiation, the original non-ion-implanted TiO_2 photocatalyst did not exhibit any photocatalytic activity.

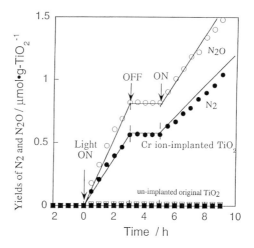

Fig. 1.10 Photocatalytic decomposition of NO into N_2O as well as N_2 and O_2 on the Cr ion-implanted TiO_2 photocatalysts under visible light irradiation ($\lambda > 450$ nm) at 295 K. Unimplanted original TiO_2 photocatalysts did showed no photocatalytic reactivity under visible light irradiation.

The action spectrum which is the light effective for photocatalytic reactions was in good agreement with the absorption spectrum shown in Fig. 1.9, indicating that only Cr ion-implanted TiO_2 photocatalysts were effective for the photocatalytic decomposition of NO. Furthermore, various photocatalytic reactions, such as the photocatalytic isomerization of *cis*-2-butene to produce *trans*-2-butene and 1-butene and the photocatalytic degradation of alcohol in water, were also possible under visible light irradiation ($\lambda > 450$ nm) at 275 K. In our findings, the metal ion-implanted TiO_2 photocatalysts enabled the absorption of visible light up to a wavelength of 550–600 nm, hence we named them 'second-generation TiO_2 photocatalysts'[23–27]

On the other hand, the photocatalytic decomposition of NO into N_2O as well as N_2 and O_2 proceeded both on the original non-ion-implanted TiO_2 and on the metal ion-implanted TiO_2 photocatalysts under UV light irradiation ($\lambda > 380$ nm). Under UV light irradiation, the photocatalytic reactivity for the decomposition of NO on the metal ion-implanted TiO_2 photocatalyst was almost the same as for the non-ion-implanted TiO_2 photocatalyst, suggesting that the implanted metal ions which are highly dispersed inside the deep bulk of TiO_2 do not work as the electron-hole recombination center but only work to modify the electronic property of TiO_2, enabling the absorption of visible light.

The maximum depth concentration of the metal ions implanted was found to be about 500–700 Å, its depth strongly depending on the voltage of the acceleration of the metal ions. Three-dimensional images obtained by detailed SIMS analysis indicated that the metal ions were deeply injected up to a depth of about 2000 Å from the surface while being highly dispersed within the catalyst. Furthermore, XPS measurements of the catalysts did not show any evidence of the presence of metal ions on the TiO_2 surface, indicating that the metal ions were highly dispersed within the deep bulk of TiO_2 and not on the surface.

1.5 Summary

The principles of photocatalysis and photocatalytic reactions have been briefly introduced as compared to photosynthesis in green plants. Photocatalysis can be considered the most important and new environmentally-friendly, clean chemical technology for the 21st century. In fact, various applications of TiO_2 photocatalysts to improve – our environment have already been introduced especially, successful developments in the purification of the polluted atmosphere as well as toxic water using ultraviolet light having a wavelength shorter than 380 nm, that is a larger energy than the bandgap energy of TiO_2 semiconductors. Furthermore, the advanced metal ion-implantation method has been successfully applied to modify the electronic properties of TiO_2 photocatalysts enabling the absorption of visible light even longer than 550 nm so that they are able to operate efficiently under visible or solar light irradiation. The design and development of such unique TiO_2 photocatalysts can be considered a breakthough in the efficient and large scale utilization of solar energy.

Acknowledgements

The author is much indebted to the Petroleum Energy Center (PEC) under the sponsorship of the New Energy and Industrial Technology Development Organization (NEDO) of the Ministry of International

Trade and Industry (MITI) of Japan and the Research Institute of Innovative Technology for the Earth (RITE) for their financial support.

References

1. Zamaraev, K.I. In *Studies in Surface Science and Catalysis*, Eds., Hightower, J.W.; Delgass, W.N.; Iglesia, E.; Bell, A.T. (Elsevier, Amsterdam), **1996**, 35.
2. Kubokawa, Y.; Anpo, M.; Honda, K.; Ito, K. In *Hikari-shokubai (Photocatalysis)*, Eds., Kubokawa, Y.; Honda, K.; and Saito, Y. (Asakura-shoten, Tokyo), **1988**.
3. Anpo, M.; Kubokawa, Y. *Res. Chem. Intermed.*, **1987**, 8, 105–148.
4. Anpo, M. *Solar Energy Materials and Solar Cells*, **1995**, 38, 221–238.
5. Anpo, M.; Yamashita, H. In *Surface Photochemistry*, Ed., Anpo, M. (John Wiley & Sons, Chichester), **1996**, 117–164.
6. Anpo, M. *Catal. Surv. Jpn.*, **1997**, 1, 169–179.
7. Serpone, N.; Pelizzetti, E. *Photocatalysis—Fundamentals and Applications* (John Wiley & Sons, London) **1989**.
8. Fujishima, A.; Hashimoto, K.; Watanabe, T. *Hikari Kurin Kakumei*, (CMC, Tokyo), **1997**.
9. Anpo, M.; Yamashita, H.; Zhang, S.G. In *Research Opportunities in Photochemical Sciences*, (NREL, Colorado), **1996**, 226.
10. Anpo, M.; Yamashita, H. In *Heterogenous Photocatalysis*, Ed., Schiavelo, M. (John Wiley & Sons, Chichester), **1997**, 133–167.
11. Anpo, M.; Matsuoka, M.; Mishima, H.; Yamashita, H. *Res. Chem. Intermed.*, **1997**, 23, 197–217.
12. Anpo, M.; Matsuoka, M.; Yamashita, H. *Catal. Today*, **1997**, 35, 177–181.
13. Yamashita, H.; Ichihashi, Y.; Anpo, Hashimoto, H.; Louis, C.; Che, M.*J. Phys. Chem.*, **1996**, 100, 16041–16046.
14. Ibusuki, T.; Takeuchi, K. *J. Mol. Catal.*, **1994**, 88, 99–102.
15. Report of the Fieldwork Test of the Photocatalytic Building Materials for the Elimination of NOx, Eds., Takami, K.; Sagawa, N. (Osaka Prefectural Government), **1996**.
16. Anpo, M.; Yamashita, H.; Ichihashi, Y.; Fujii, Y.; Honda, M. *J. Phys. Chem.*, **1997**, 101, 2632–2636.
17. Yamashita, H.; Fujii, Y.; Ichihashi, Y.; Zhang, S.G.; Ikeue, K.; Park, D.R.; Koyanao, K.; Tatsumi, T.; Anpo, M. *Catal. Today*, **1998**, 45, 221–227.

18. Zhang, S.G.; Fujii, Y.; Yamashita, H.; Koyano, K.; Tatsumi, K.; Anpo, M. *Chem. Lett.*, **1997**, 659–660.
19. Anpo, M.; Yamashita, H.; Ikeue, K.; Fujii, Y.; Zhang, S.G.; Ichihashi, Y.; Park, D.R.; Suzuki, Y.; Koyanao, K.; Yatsumi, T. *Catal. Today*, **1998**, 44, 327–332.
20. Sayama, K.; Arakwa, H. *J.C.S., Faraday Trans.*, **1997**, 93, 1647–1651.
21. Domen, K. In *Surface Photochemistry*, Ed., Anpo, M. (John Wiley & Sons, Chichester), **1996**, 1–18.
22. Moon, S.C.; Mametsuka, H.; Suzuki, E.; Anpo, M. *Chem. Lett.* **1998**, 447–448.
23. Anpo, M.; Ichihashi, Y.; Yamashita, H. *Optronics*, **1997**, 6, 161–165.
24. Yamashita, H.; Anpo, M. *Chemistry* (Kagaku), **1997**, 52, 74–75.
25. Anpo, M.; Yamashita, H.; Ichihashi, Y. *ULVAC Tech. J.*, **1997**, 47, 15–19.
26. Patent, No. 9–262482, *Photocatalyst, Method of Producing the Photocatalyst, and Photocatalytic Reaction Method*, USA, Germany, England, Italy, **1997**.
27. Anpo, M.; Ichihashi, Y.; Takeuchi, M.; Yamashita, H. *Res. Chem. Intermed.* **1998**, 24, 143–149.

2 'Green oxidations': horseradish peroxidase (HRP)-catalyzed regio and diastereoselective preparation of dilignols

Ezio Bolzacchini, Simone Meinardi, Marco Orlandi,
Bruno Rindone, Gösta Brunow, Pekka Pietikainen, and
Petteri Rummakko

2.1 Introduction

The selective oxidation of organic compounds at room temperature is one of the more important objectives of synthetic chemistry. A high valent metal ion such as Fe(III), Cu(II), Ce(IV) or Pb(IV) as the stoichiometric reagent or as the catalyst is often required.[1] The control of the metal ion is sometimes difficult, and requires expensive treatments or recycling procedures. Recent attempts of a metal-free approach have used metal-centered systems which mimic monooxygenases. Examples are cytochrome P-450 and methanemonooxygenase mimics.[2] An alternative is the use of dioxiranes as oxidizing agents.[3] Ozone has also been used for the oxidation of cycloalkanes,[4] ethers,[5] and amines.[6] 'Green chemistry' alternatives for oxidative methods are still needed. An enzyme-catalyzed oxidative synthetic method has the advantage of mild reaction conditions, fast reaction rates and easy control of the reactants.[7]

Phenol oxidative coupling is a key reaction in the formation of many natural products.[8] It has been estimated that over 2000 known alkaloids can be obtained by oxidative coupling of phenolic precursors, and this synthetic method is widely used because of the biological importance of many of them.[9] The use of enzyme-catalyzed phenol oxidative coupling reactions has been one of the goals of chemistry for many years.[10,11]

The use of horseradish peroxidase (HRP) for the removal of phenols in waste water has been proposed recently.[12] This enzyme has been also used for the enantioselective oxidation of 2-naphthols to 1,1′-binaphthyl-2,2′-diols[13] and for the oxidation of 3-t-butyl-4-hydroxyanisole.[14] Furthermore, HRP-catalyzed oxidative polymerization of o-phenylendiamine gives a soluble polymer having an iminophenylene structure with a molecular weight of 2×10^4.[15]

The use of enzymes as catalysts in oxidation reactions is biomimetic since the biosynthesis of lignin from phenolic phenylpropenoids occurs via oxidative processes catalyzed by enzymes such as laccase, which convert the phenols into phenoxy radicals by electron abstraction.[16] These radicals undergo dimerization via carbon–carbon and carbon–oxygen bond formation.[17] In this step, some stereogenic carbons are formed. Mimicking this process could have synthetic value since some dilignols have important biological activity.[18]

2.2 The HRP-catalyzed oxidative phenol coupling with alkylphenols

2.2.1 4-methylphenol (1)

The HRP-catalyzed oxidation of 4-methylphenol (1), 10 mM, with hydrogen peroxide 10 mM in 0.1 M phosphate buffer pH = 7, was performed for different substrate to catalyst ratios and for different reaction times. The main product was always the carbon–carbon and carbon–oxygen dimer Pummerer's ketone (2) (Fig. 2.1) (4a,9b-dihydro-8,9b-dimethyl-3(4H)-dibenzofuranone). Two minor reaction products were the carbon–carbon dimer 2,2′-dihydroxy-5,5′-dimethylbiphenyl, (3) and the carbon–carbon trimer 2,2′,2′-trihydroxy-5,5′5″-trimethyl-m-terphenyl (4). The selectivity in Pummerer's ketone was fairly good for short reaction times.[19] Yields and conversions were determined by HPLC and are shown in Table 2.1. Pummerer's ketone (2) had also been obtained in poor yield in the oxidation of 4-methylphenol (1) in a mixture of homogenized potato peelings and hydrogen peroxide.[20]

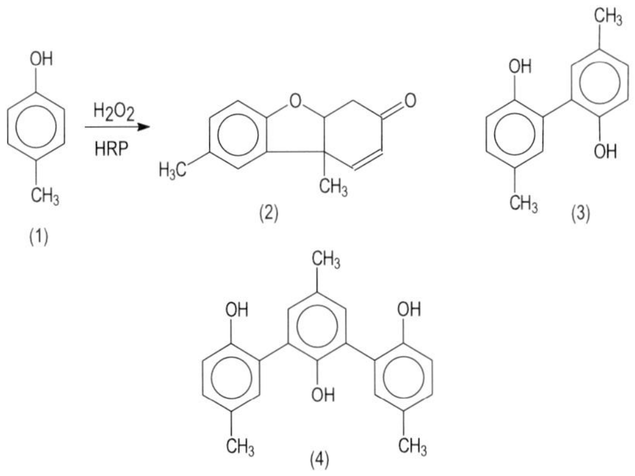

Fig. 2.1 Scheme of the HRP-catalyzed oxidation of 4-methylphenol.

Table 2.1 Conversion and product distribution % in the HRP-catalyzed oxidation of 4-methylphenol (1) with hydrogen peroxide.

[HRP] (mg ml^{-1})	Time (h)	Conversion (%)	(2) (%)	(3 + 4) (%)
0.10	0.5	98	12	3
0.02	19.0	90	19	7

2.2.2 4-terbutylphenol (5)

The HRP-catalyzed oxidation of 4-terbutylphenol (5), 10 mM, with hydrogen peroxide, 10 mM, was carried out under a variety of conditions. The carbon–carbon dimer 2,2′-dihydroxy-4,4′-di-t-butyldiphenyl (6) and the carbon–oxygen dimer 2-hydroxy-4,4′di-t-butyldiphenyl oxide (7) were formed (Fig. 2.2). The selectivity in this reaction did not depend on the reaction time but was correlated with the buffer's pH and the amount of methanol added as organic cosolvent into the aqueous solution. The highest selectivity was obtained at pH = 3 in a solution of 80% aqueous buffer, 20% methanol. Yields and conversions were determined by HPLC and are reported in Table 2.2.

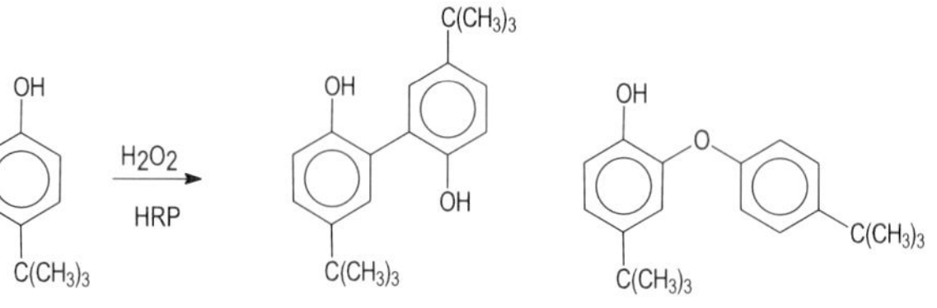

Fig. 2.2 Scheme of the HRP-catalyzed oxidation of 4-terbutylphenol.

Table 2.2 Conversion and product distribution % in the HRP-catalyzed oxidation of 4-t-butylphenol (5) with hydrogen peroxide for different amounts (%) of methanol in the aqueous buffer.

pH	Methanol (%)	Time (h)	Conversion (%)	(6) + (7) (%)
3	20	0.5	93	34
3	20	1.0	93	36
6	20	0.5	91	12
6	20	1.0	90	9
3	0	1.0	96	6
6	50	1.0	99	5
3	90	1.0	11	–

2.2.3 2,6-dimethylphenol (8)

2,6-dimethylphenol (8) gave a 70% yield of the carbon–carbon dimer (9) in the HRP-catalyzed oxidation with hydrogen peroxide at pH 3 or pH 6 (Fig. 2.3). This quinoide product is similar to that obtained in 90% yield

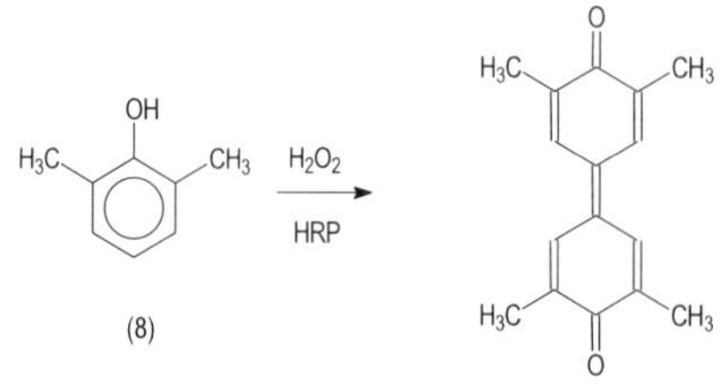

Fig. 2.3 Scheme of the HRP-catalyzed oxidation of 2,6-dimethylphenol.

from the dimerization of 2,6-dimethoxyphenol with an enzyme isolated from *Polyporus versicolor*.[21]

2.3 The HRP-catalyzed phenol coupling with alkenylphenols

2.3.1 The enzyme–substrate complex

The study of the enzyme–substrate complex in HRP-catalyzed reaction was performed using phenylpropenoidic compounds as substrates. Previous work had shown[22] that the enzyme in the native form interacts with aromatic donors and forms 1 : 1 complexes. Moreover, the addition of the substrate causes very little change in the electronic spectrum of HRP. This indicates that the binding site is quite far from the ferric center. Further work was performed using NMR techniques.[23]

The influence of an R group linked to the vinyl chain in compounds (10–13) (Fig. 2.4) was studied by measuring the binding constants of the compounds to HRP. A spectrophotometric titration allowed the binding constants K_b to be obtained. Enzyme–substrate binding measurements were performed by adding small amounts of concentrated solutions of the substrates (0.1 M in MeOH) to the enzyme solution (10^{-5} M in 1 : 1 aqueous buffer 7.2 : MeOH). The same amount of substrate was added to a reference cell containing the solvent. This allowed the subtraction of the absorption of the substrate, which is close to the Soret region. All the

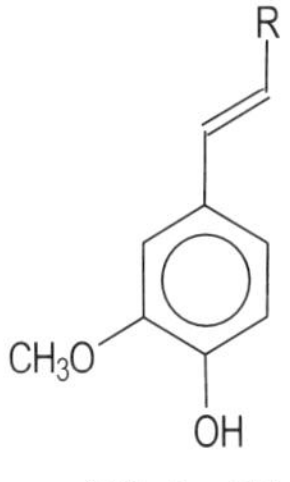

(10): R = CH$_2$OH
(11): R = COOCH$_3$
(12): R = COOH
(13): R = CH$_3$

Fig. 2.4 Four alkenylphenols used to measure the binding constants to HRP.

Table 2.3 Binding constant, K_B, and number of the sites of binding, n, for the interaction of four phenylpropenoidic compounds with HRP.

Substrate	K_B	n
E-Coniferyl alcohol (10)	0.350	1.02
E-Methyl ferulate (11)	0.157	1.03
E-Ferulic acid (12)	0.286	1.08
E-Isoeugenol (13)	0.492	1.06

spectra were then corrected for the small variation in volume. The difference spectra were obtained by subtracting the spectra of HRP from those of the enzyme–substrate complexes. The binding constants were evaluated by a least square fitting of the data using the following expression:

$$1/\Delta A = [1/(K_b \Delta A_x)]1/[S]^\circ + 1/\Delta A_x$$

where ΔA is the variation in absorbance of the enzyme caused by the addition of the substrate, ΔA_x the variation in absorbance caused by the complete formation of the enzyme–substrate complex, K_B is the binding constant and $[S]^\circ$ is the ligand concentration. The Hill plot,[24] gave the number n of the binding sites (Table 2.3). These K_b were very similar to those obtained by other authors for simpler phenolic compounds[25,26] and this suggests that the binding site for phenolic phenylpropenoids is the same found for simpler phenols and involves a tyrosine residue.[27] The similarity of the binding constants suggests a very similar reactivity for these substrates.

2.3.2 Substrate specificity

Substrate specificity of some peroxidases and the pH dependence of the oxidative coupling with hydrogen peroxide was studied using E-methyl ferulate (11) as the substrate. The conversion was always 100% for the reaction catalyzed either by lactoperoxidase (LP) from bovine milk or by HRP. On the contrary, manganese peroxidase, MnP, from the white rod fungus *Phlebia radiata* and lignin peroxidase, LiP, were less effective catalysts. The optimal pH was different for the different enzymes, but it was always lower than pH 6. Figure 2.5 shows the results thus obtained.

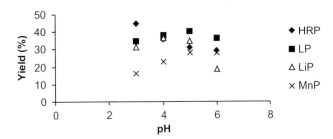

Fig. 2.5 Substrate specificity of some peroxidases and the pH dependence of the reaction of methyl ferulate (11). HRP = horseradish peroxidase, LP = lactoperoxidase (LP), MnP = manganese peroxidase, LiP = lignin peroxidase.

2.4 The HRP-catalyzed diastereoselective synthesis of phenylcoumaran dimers

2.4.1 The optimization of the synthesis

The HRP-catalyzed oxidative dimerization of E-methyl ferulate (11) with hydrogen peroxide had been shown to give the racemic *trans* phenylcoumaran (14–15).[28] The reaction was optimized by us using a mixture of aqueous buffer and organic cosolvent as reaction medium. Table 2.4 shows the results thus obtained. The enzyme was eventually immobilized by adsorption on a solid support (Celite). In organic solvents, the conversion was generally lower and more enzyme was needed for a successful transformation. Also, using methanol as organic

Table 2.4 Conversion and yield % of the phenylcoumaran dimer (14 + 15) depending on the solvent, the amount of catalyst, water (%) and the pH.

Solvent	Celite	Water (%)	HRP	pH	Conversion (%)	(14 + 15) (%)
Isopropanol	–	30–90	4.3	6	100	15
Acetone	–	30–90	4.3	6	100	15–20
Dioxane	–	90	4.3	3	100	45
Methanol	–	90	4.3	3	100	50
Ethyl acetate	+	1.0	20	//	78	42
Butyl acetate	+	1.0	20	//	60	32
Chloroform	+	0.6	20	//	81	40

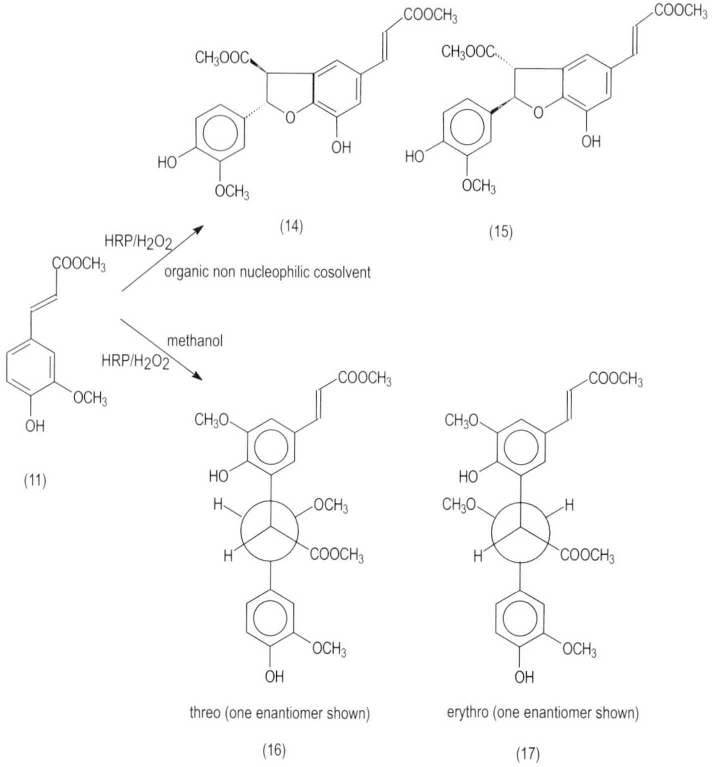

Fig. 2.6 Scheme of the HRP-catalyzed oxidation of E-methyl ferulate with organic cosolvent.

cosolvent, a nearly equimolecular mixture of the racemic threo and erythro β-O-4 dimers (16–17) were formed (Fig. 2.6).

Hence, the highest selectivity in phenylcoumarans was obtained using the solution aqueous buffer pH 3–dioxane 30%. Under these conditions 80% of the *trans*-phenylcoumaran (14–15) was obtained with a very high diastereoselection. No enantioselection resulted, as judged by its CD spectrum and by chiral HPLC analysis of the reaction mixture. This indicated that HRP did not induce enantioselectivity in these reactions. This was not unexpected since the dimerization of these phenylpropenoidic phenols is due to the formation of a phenoxy radical.[29,30]

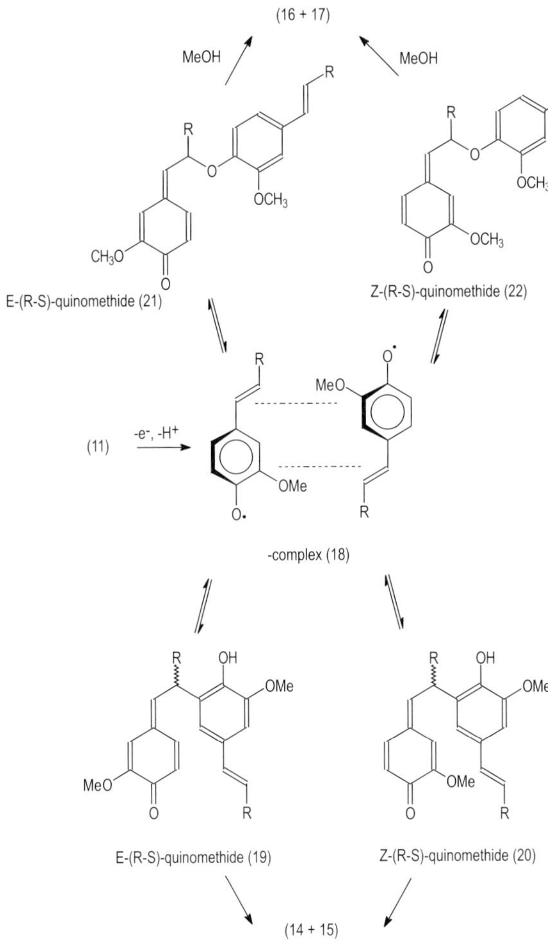

Fig. 2.7 Scheme of the postulated mechanism of the dimerization of E-methyl ferulate.

Phenoxy radicals are believed to be very persistent,[1] and the dimerization reaction is likely to be a solution process. Figure 2.7 shows the postulated mechanism. The intermediate, a π-complex (18), undergoes reversible carbon–carbon bond formation to give a group of isomeric quinomethides (19–20) having E–Z stereochemistry at the double bond and R–S chirality at the stereogenic carbon. They are two pairs of enantiomers, i.e. E, R–E, S and Z, R–Z, S. A suitable conformation of these undergoes nucleophilic attack of the phenolic oxygen to the double bond giving the phenylcoumaran dimer R,S and S,R (14 + 15).[31]

2.4.2 Molecular dynamics calculations

The control of the stereochemistry of radical reactions has been an important point in the last few years.[32–34] Concerning the diastereoselection observed in the formation of phenylcoumarans, the conformational analysis on rotating the carbon–carbon bond between the quinomethide double bond and the phenolic moiety was performed using MM2 and semiempirical calculations with UHF-AM1. In all the quinomethides (19–20), the quinomethide ring and the phenolic ring were twisted approximately 110°, thus generating two conformations (conformation 1 and 2) of different stability (Table 2.5). Only the less stable conformation (conformation 2) of the isomers Z,R and E,R could generate the *cis*-phenylcoumaran. The *trans*-phenylcoumaran was formed in all other cases. Figure 2.8 shows the shape of quinomethide E,R conformation 1 which gives the *trans*-phenylcoumaran E,R. Figure 2.9 shows the shape of quinomethide E,S conformation 2 which gives the *cis*-phenylcoumaran R,R. Hence, the diastereoselection was likely to derive from a thermodynamic control in the equilibrium

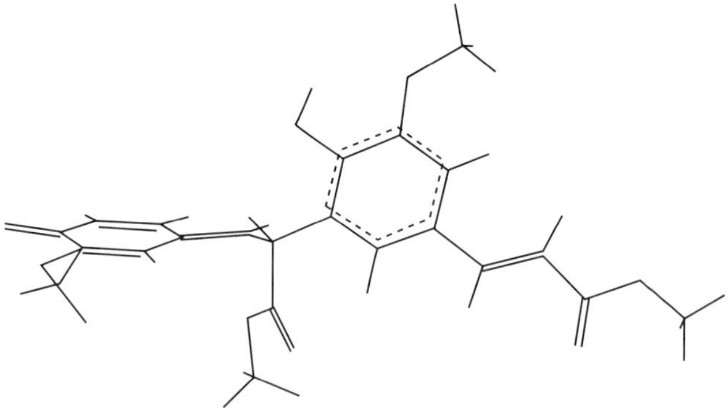

Fig. 2.8 Quinomethide E,R conformation 1 which gives the *trans*-phenylcoumaran S, R.

Table 2.5 Calculated stability of two conformations of the four quinomethides (19–20) and stereochemistry of the resulting phenylcoumaran.

Configuration	MM2 energy (kcal mol⁻¹)	AM1 heat of formation (kcal mol⁻¹)	Conformation	Distance carbon–oxygen (Å)	Dihedral angle between rings (Degrees)	Phenylcoumaran formed
Z,R	18.47	−231.00	1	3.51	108.10	*Trans* S,R
Z,R	29.45	−225.56	2	3.07	111.54	*Cis* R,R
Z,S	19.01	−228.81	1	2.97	110.93	*Trans* R,S
Z,S	31.45	−219.67	2	3.39	110.81	*Trans* R,S
E,R	21.50	−232.68	1	3.57	111.41	*Trans* S,R
E,R	25.72	−224.51	2	3.13	111.68	*Cis* R,R
E,S	21.93	−229.51	1	3.04	110.20	*Trans* R,S
E,S	35.37	−221.14	2	3.46	110.69	*Trans* R,S

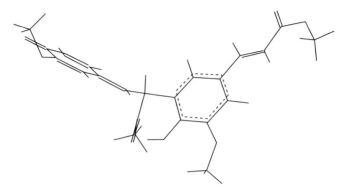

Fig. 2.9 Quinomethide E,R conformation 2 which gives the *cis*-phenylcoumaran R,R.

formation of these conformers from the original π-complex (18) and the consequent control of the following nucleophilic attack of the phenolic hydroxyl group on the quinomethide double bond. Here, the non-bonded distance was slightly more than 3 Å. To achieve further confirmation on this point, the shape of the two conformations of the four isomeric quinomethides (19–20) in water was investigated by calculations of molecular dynamics. The geometries were first optimized using semiempirical calculations (UHF-AM1), then included in a water box 20 × 20 × 30 Å and recalculated. The conformations of minimum energy in water again indicated a preferential attack of the phenolic group to give the racemic *trans*-phenylcoumaran (14–15).

Fig. 2.6 shows also the mode of formation of the β-O-4 dimers (16–17). Here, another group of isomeric quinomethides (21) and (22), having E–Z stereochemistry of the double bond and the R–S chirality at the stereogenic carbon, undergo nucleophilic attack by methanol forming the reaction products. Here, no control of the attack of methanol on the intermediate quinomethide was possible, and no diastereoselection resulted.

In conclusion, the HRP-catalyzed oxidative coupling of phenols with hydrogen peroxide is environmentally by friendly since the reaction occurs in water at room temperature and the wastes do not contain metals. This methodology meets the requirements of 'green chemistry' and may be used in cases in which pollution prevention is important.

Moreover, it is in line with the 'industrial ecology' concept which gives priority to environmentally by friendly procedures applied to renewable resources. Polyphenols from wood belong to this family of organic compounds. The selectivity obtained in this synthetic procedure is an additional value since diastereoselection is obtained in some cases. The diastereolectivity may be anticipated by monitoring at the reaction mechanism using molecular mechanics and molecular dynamics calculations.

Acknowledgement

This work was supported by a grant of the Italian Ministry for Scientific Research and by the Technology Development Center of Finland (TEKES) and the Academy of Finland.

References

1. Sheldon, R.A.; Kochi, J.K. *Metal-Catalyzed Oxidation of Organic Compounds*. Academic Press, New York, **1981**.
2. Aewnds, I.W.C.E.; Mac Faul, P.A.; Snelgrove, D.W.; Wayer, D.D.M.; Ingold, K.U. In Minisci, F. ed., *Free Radicals in Biology and Environment*. Kluver Academic Publishers, New York, **1997**, p. 79–90.
3. Reiser, O. *Angew. Chem. Int. Ed.* **1994**, *33*, 69–72; Bravo, A.; Fontana, F.; Fronza, G.; Mele, A.; Minisci, F. *J. Chem. Soc. Chem. Commun.,* **1995**, 1573–1574; Vanni, R.; Garden, S.J.; Banks, J.; T. Ingold K.U., *Tetrahedron Lett*, **1995**, 7999–8002; Adam, W.; Curci, R.; D'Accolti, L.; Dinoi, A.; Fusco, C.; Gasparrini, F.; Kluge, R.; Paredes, R.; Schulz, M.; Smerz, A.K.; Veloza, L.A.; Weinkotz, S.; Winde, R. *Chemistry – A European Journal*, **1997**, *3*, 105–10.
4. Bailey, P.S. *Ozonation in Organic Chemistry*. Academic Press, New York, **1981**, p. 255.
5. Brambilla, A.; Bolzacchini, E.; Orlandi, M.; Polesello, S.; Rindone, B. *Water Research*, **1997**, *31*, 1839–1846.
6. Bolzacchini, E.; Brambilla, A.; Orlandi, M.; Polesello, S.; Rindone, B. *Wa. Sci. Tech*, **1994**, *30*, 129–136.
7. Setala, H.; Pajunen, A.; Kilpelainen, I.; Brunow, G. *J. Chem. Soc. Perkin Trans.* **1994**, *1*, 1163–1165.

8. Guo, Z.; Salamonczyk, G.M.; Han, K.; Machiya, K.; Sih, C.J. *J. Org. Chem.*, **1997**, *62*, 6700–6701.
9. Scott, A.I. *Quart. Rev.*, **1965**, *19*, 1–35.
10. Brown, B.R.; Bocks, S.M. *Enzyme chemistry of phenolic compounds.* ed. J.B. Pridham. Pergamon Press, Oxford, **1963**.
11. Holland, H.L. *Organic synthesis with oxidative enzymes.* VCH Publisher, Inc., New York, **1992**.
12. Trotta, F.; Ferrari, R.P.; Laurenti, E.; Moraglio, G.; Trossi, A. *J. Inclusion. Phenom. Mol. Recogn.*, **1996**, *1–3*, 225–228.
13. Sridhar, M.; Vadivel, S.K.; Bhalerao, U.T. *Tetrahedron Lett.*, **1997**, 5695–5696.
14. Tajima, K.; Hashizaki, M.; Yamamoto, K.; Mizutani T. *Drug Metab. Dispos.*, **1992**, *20*, 816–820.
15. Kobayashi, S.; Kanakeko, I.; Uyama, H. *Chem. Lett.*, **1992**, *3*, 393–394.
16. McDonald, P.D.; Hamilton, G.A. *Mechanism of phenolic oxidative coupling reactions.* In *Oxidation in organic chemistry*, ed. W.S. Trahanovsky. Academic Press, New York, **1973**, pp. 97–134.
17. Neish, A.C. *Monomeric intermediates in the biosynthesis of lignin and related compounds: the formation of wood in forest trees.* Academic Press, New York, **1968**, pp. 219–239.
18. Ayres, D.C.; Loike, J.D. *Lignans – Chemicals, biological and clinical properties.* Cambridge University Press, Cambridge, **1990**, pp. 12–84.
19. Pietikainen, P.; Adlercreutz, P. *Appl. Microb. Biotechnol.*, **1990**, *33*, 455–458.
20. Kametani, T.; Takano, S.; Kobari, T. *J. Chem. Soc. C*, **1969**, 9–12.
21. Eriksson, H.; Gustafsson, J.A.; and Sjovall, J. *Eur J. Biochem*, **1968**, *6*, 219–225.
22. Schejter, A.; Laniz, A.; Epstein, N. *Arch. Biochem. Biophys.*, **1976**, *174*, 36–44.
23. Banci, L.; Bertini, I.; Bini, T.Z.; Tien, M.; Turano, P. *Biochemistry*, **1993**, *32*, 5825–5831.
24. Casella, L.; Gullotti, M.; Poli, S.; Selvaggini, C.; Beringhelli, T.; Marchesini, A. *Biochem.*, **1994**, *33*, 6377–6386.
25. Schonbaum, G.R.; Tavares de Sousa, J. *J. Biol. Chem.*, **1973**, *248*, 502–511.
26. Paul, K.G.; Ohlsonn, P.I. *Acta Chem. Scand. Ser. B*, **1978**, *32*, 395–404.
27. Sakuradaj, J.; Takahashi, S.; Hosoyai, T. *J. Biol. Chem.,* **1986**, *261*, 9657–9662.
28. Katayama, Y.; Fukuzumi, T. *Mokuzay Gakkaishi*, **1978**, *24*, 664–667.

29. Fengel, D.; Wenger, G. *Wood, chemistry, ultrastructure reactions*. Walter de Gruyster ed. Co, Berlin, **1994**.
30. Doerge, D.R.; Taurog, A.; Dorris, M.L. *Arch Biochem Biophys*, **1994**, *315*, 90–99.
31. Chioccara, F.; Poli, S.; Rindone, B.; Pilati, T.; Brunow, G.; Pietikäinen, P.; Setälä, H. *Acta Chem. Scand.*, **1993**, *47*, 610–616.
32. Kim, B.H.; Curan, D.P. *Tetrahedron*, **1993**, *49*, 293–318.
33. Porter, N.A.; Giese, B.; Curran, D.P. *Acc. Chem. Res*, **1991**, *24*, 296–304.
34. Smadja, W. *Synlett.*, **1994**, 1–26.

3 *Fine chemicals preparation* via *solventless reactions under heterogeneous catalysis*

Franca Bigi, Raimondo Maggi, and Giovanni Sartori

3.0 Introduction

The manufacture of fine chemicals by safe and environmentally friendly strategies represents the fundamental challenge of industrial and synthetic chemistry at the beginning of the 21st century.[1] The application of cleaner heterogeneous catalysts under solventless conditions, or with small amounts of environmentally compatible solvents, in highly selective processes would allow the minimization of inorganic as well as organic waste production.[2] Our own research interests have recently focused on the use of the heterogeneous catalysis to promote selective synthesis of fine chemicals. Here we provide several examples which indicate that the combination of the 'one-pot' strategy[3] with the use of solid acid catalysts allows the preparation of specific target compounds with minimum production of pollutants and reduction of the costs, according to the increasingly stringent environmental regulations.

3.1 Reaction of aromatic amines with ethyl acetoacetate: solventless synthesis of ureas

The first example concerns the reaction of amines and ethyl acetoacetate for the preparation of *N,N'*-disubstituted ureas **5** and **6** under solventless conditions. Substituted ureas represent a class of commercially important compounds utilized in different areas of industrial chemistry ranging from antioxidants, additives for detergents, inhibitors of corrosion, herbicides and tranquillizers.[4] Typical methods for the manufacture of ureas are essentially based on phosgene and isocyanates[5] or phosgene substitutes.[6] Unfortunately these efficient reagents are toxic, expensive to handle and generate large amounts of waste, mainly due to the work-up of chloride by-products.

3.1.1 Synthesis of *N,N'*-diphenylureas[7]

N,N'-Diphenylureas **5** (DPUs) could be obtained as the sole reaction products by heating a mixture of aromatic amines and ethyl acetoacetate in the presence of zeolite HSZ-360 (Fig. 3.1).[8] The reaction of aromatic amines with ethyl acetoacetate to produce β-arylaminocrotonic acid ethyl esters **7** or acetoacetoanilides **3** depending on the temperature has been described early in the literature (Fig. 3.2).

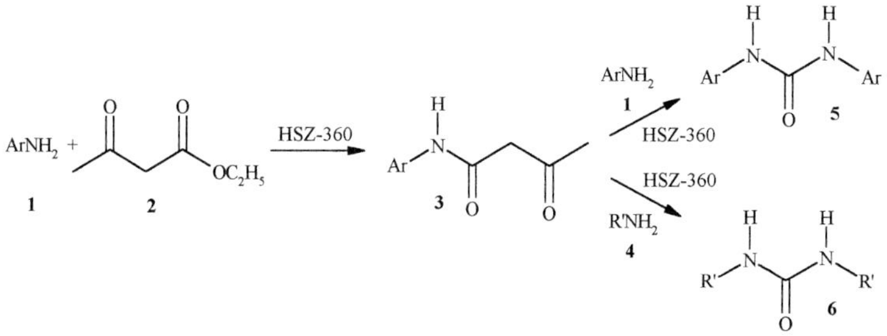

Fig. 3.1 Synthesis of *N,N'*-disubstituted ureas: general scheme.

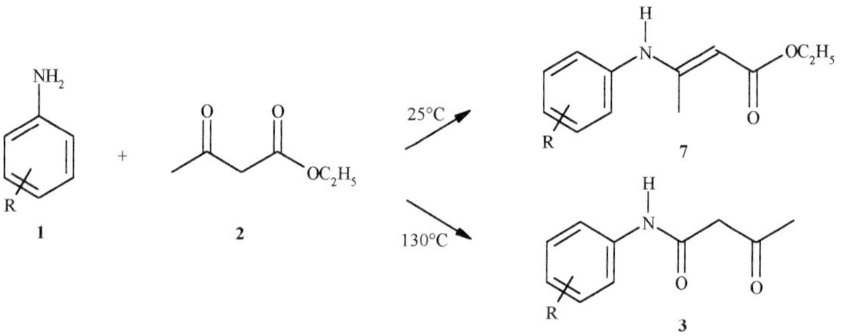

Fig. 3.2 Uncatalyzed reaction of aromatic amines and ethyl acetoacetate: temperature effect.

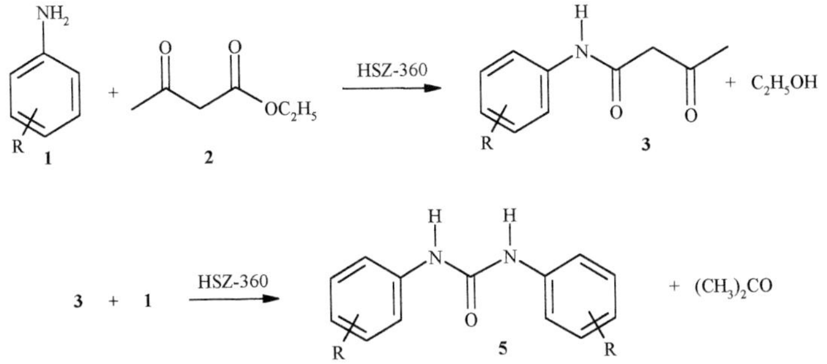

Fig. 3.3 Synthesis of symmetric N,N'-diarylureas: reaction pathway.

By again using aromatic amines and ethyl acetoacetate as the starting materials, but in the presence of zeolite HSZ-360, diphenylureas could be obtained under solventless conditions through a completely different chemical behavior (Fig. 3.3). The formation of compound **5** could be attributed to the initial production of acetoacetanilide **3** and its subsequent reaction with a second molecule of aromatic amine **1** to give DPU and acetone. This hypothesis was, in part, confirmed by quantitative production of diphenylurea by heating a 1 : 1 mixture of acetoacetanilide and aniline in the presence of zeolite HSZ-360 (Fig. 3.3). The methodology could be extended to different aromatic amines. Some synthetic results are reported in Table 3.1.

A typical experimental procedure is as follows: a flask containing a mixture of the selected aromatic amine (10 mmol) and zeolite HSZ-360 (0.5 g) was put in a hot oil bath (180 °C) and ethyl acetoacetate (0.8 g, 0.8 ml, 6 mmol) was added dropwise during 1 min. The mixture was efficiently stirred at the same temperature for 5 h. Boiling methanol

Table 3.1 Synthesis of variously substituted N,N'-diphenylureas 5.

Entry	R	Yield (%)	Selectivity (%)
a	4-OCH$_3$	76	95
b	H	66	95
c	3-CH$_3$	77	93
d	4-CH$_3$	73	94
e	4-Cl	58	96

containing 5% of water (150 ml) was added to the reaction mixture cooled to 100 °C. After filtration the product was recovered from the solution by addition of more water and cooling. In place of methanol, hot DMSO could be successfully utilized under the same conditions.

It is noteworthy that the activity of HSZ-360, recovered by filtration, washed with acetone and reactivated by heating at 500 °C for 8 h, was the same for five runs.

3.1.2 Synthesis of N,N'-dialkylureas

On the basis of the above shown mechanistic hypothesis, N,N'-disubstituted alkylureas were synthesized by reacting acetoacetoanilide **3** with an excess of aliphatic amines. In fact, acetoacetanilide **3** reacts twice and the first products **8** compete for the reagents RNH_2. Dialkylureas **6** were isolated by utilizing longer reaction times (3 hours) (Fig. 3.4).

The direct reaction of aliphatic amines with ethyl acetoacetate gave the corresponding β-alkylaminocrotonic acid ethyl ester in quantitative yield. The different chemical behavior of aliphatic amines compared to the aromatic ones toward ethyl acetoacetate can be attributed to the higher nucleophilicity of the former substrates that is responsible for the attack on the harder keto group. N,N'-Dialkylureas **6** were thus obtained by reacting acetoacetoanilide with aliphatic amines at 180 °C for 3 hours under solventless conditions in the presence of the zeolite HSZ-360. Synthetic results are reported in Table 3.2.

A typical experimental procedure is as follows: a flask containing a mixture of the selected aliphatic amine (40 mmol), acetoacetoanilide (1.8 g, 10 mmol) and zeolite HSZ-360 (0.5 g) was put in a hot oil bath (180 °C) and an efficient stirring was continued for 5 h. After cooling to room temperature, the slurry was washed with boiling methanol containing 5% water (2 × 100 ml). After filtration the product was recovered from the solution by addition of more water and cooling. In

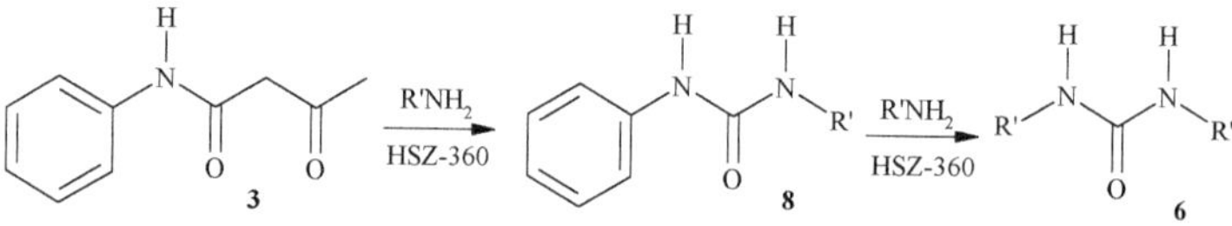

Fig. 3.4 Synthesis of symmetric N,N'-dialkylureas: reaction pathway.

Table 3.2 Synthesis of variously substituted N,N'-dialkylureas 6.

Entry	R'	Yield (%)	Selectivity (%)
a	PhCH$_2$	95	97
b	C$_6$H$_{11}$	85	95
c	C$_8$H$_{17}$	82	93
d	C$_{10}$H$_{21}$	85	95
e	(R)–Ph(CH$_3$)CH	75	97

place of methanol, hot DMSO could be successfully utilized under the same conditions.

3.2 Reaction of aromatic substrates with unsaturated electrophiles: synthesis of 1,1-diarylethylenes, *ortho-iso*pentenylphenols, chromanes and chromenes

Earlier studies on the 'metal-template electrophilic substitution' of ambidental aromatic substrates have confirmed that *ortho*-regioselective alkylation of phenols with phenylacetylene and conjugated dienes could be performed under homogeneous catalysis in the presence of a stoichiometric amount of Lewis acids.[10] Both processes were recently revised under heterogeneous acid catalysis.

Results depicted in Fig. 3.5 indicate that phenylacetylene, enynes and isoprene react under solid acid catalysis with aromatic substrates affording different classes of fine chemicals. All reactions were performed under solventless conditions or in *ortho*-dichlorobenzene (or decaline).[11]

3.2.1 Synthesis of 1,1-diarylethylenes[12]

The alkylation of aromatic hydrocarbons and phenols with phenylacetylene was performed as follows: 1.0 g of zeolite HSZ-360 was added to a solution of phenylacetylene (1.0 g, 10 mmol) and the selected aromatic compound (10 mmol) in *ortho*-dichlorobenzene (or decaline) (6 ml). The mixture was heated at 110 °C under stirring for 14 h. After cooling to room temperature, filtration and distillation of the solvent, products were purified by flash chromatography (eluant hexane/ethyl acetate: 90/10). Benzene and *meta*-xylene were utilized in excess as solvent reagents.

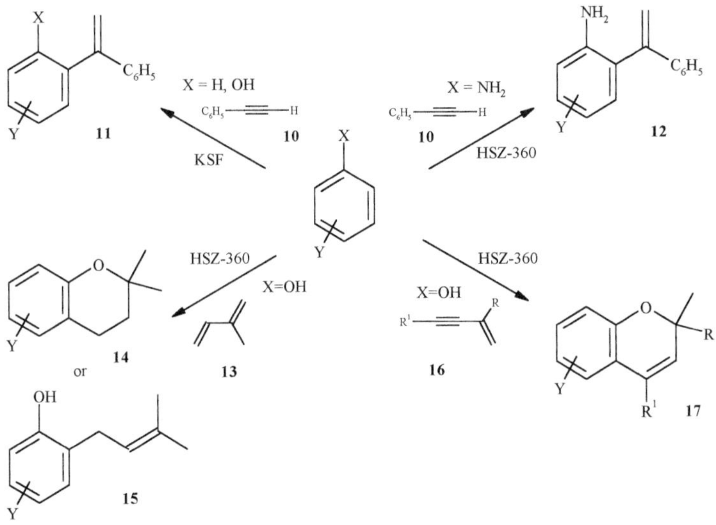

Fig. 3.5 Reaction of aromatic substrates with unsaturated electrophiles: general scheme.

As clearly emerges from synthetic results reported in Fig. 3.6, yields and selectivities of products **11** are satisfactory or excellent. Competitive isomerization and transalkylation were never observed in the reaction with *meta*-xylene (entry b); moreover, only products of monoalkylation were detected in all cases. Concerning the absolute *ortho*-regioselective control observed with phenol (entry d), the product is likely to have been formed through the intervention of an hydrogen bond between the hydroxy group and the alkyne on the analogy of the *ortho*-regioselective alkylation of phenols with alkenes promoted by aluminium phenolates.[13] In fact, when anisole was reacted under the above conditions, a 1 : 1 mixture of *ortho*- and *para*-phenylethenylanisoles was obtained.

With regard to the phenylacetylene, variable amounts of acetophenone (5–20%) were detected in all experiments due to the competitive reaction with traces of water linked to the catalyst.[14] On the other hand, attempted use of a completely dried catalyst resulted in a remarkable lowering of reactivity. Indeed 1-phenyl-1-(2-hydroxyphenyl)ethylene was obtained in 10–20% yield by using HSZ-360 previously heated at 500 °C for 10 h or heated under high vacuum at 280 °C for 10 h.

Entry	Substrate	Conversion (%)	Product	Yield (%)	Selectivity (%)
a	(benzene)	49	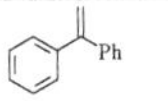 (1-phenylethenyl)benzene, Ph	40	82
b	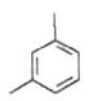 (toluene)	94	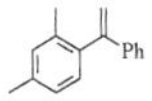 methyl-substituted styrene, Ph	80	85
c	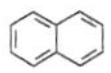 (naphthalene)	80	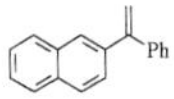 2-(1-phenylethenyl)naphthalene, Ph	70	88
d	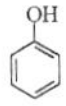 (phenol, OH)	55	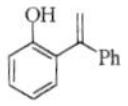 2-(1-phenylethenyl)phenol, OH, Ph	52	95
e	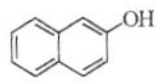 (2-naphthol, OH)	93	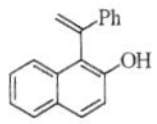 Ph, OH substituted naphthalene	90	97
f	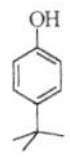 (4-tert-butylphenol, OH)	74	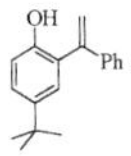 4-tert-butyl-2-(1-phenylethenyl)phenol, OH, Ph	70	95

Fig. 3.6 Synthesis of various diarylethylenes **II**.

Additional experiments demonstrated that the zeolite, recovered by Büchner filtration and washed with diethy ether, could be reused six times without loss of activity to perform the reaction with *para-tert*-butylphenol.

Concerning the interaction with the catalyst, all reagents could diffuse through the pores of the zeolite HSZ-360, which have a dimension of 7.4 Å; however, particularly bulky products, such as compound **11e** (Fig. 3.6 entry e), would give rise to more difficult intracrystalline diffusion. Results from Fig. 3.6 seem to suggest that the intrinsic reactivity of the aromatic substrates represents a factor more important than the molecular sieving effect of the zeolite. Nevertheless, the crystallinity of the catalyst plays a crucial role in the activation of reactants. In fact, by reacting the phenol with phenylacetylene in the presence of similar acidic but amorphous silica-alumina,[15] 1-phenyl-1-(2-hydroxyphenyl)ethylene was obtained in lower yield (28%), but with the same *ortho*-regioselectivity.

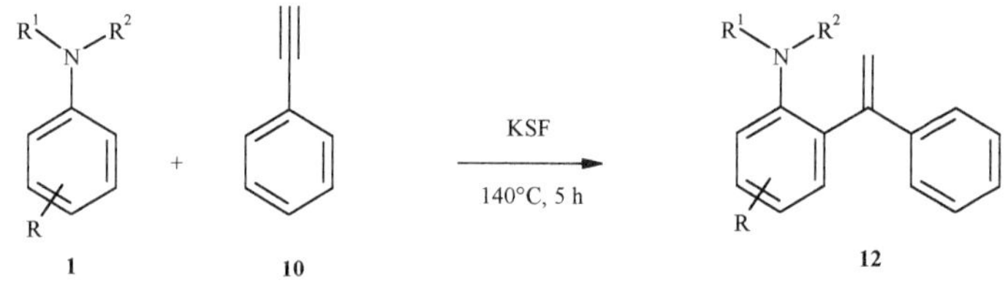

Fig. 3.7 Synthesis of *ortho*-aminodiarylethylenes.

A similar methodology was successfully applied to the synthesis of *ortho*-styrylanilines that represent useful synthons for the preparation of biologically active dihydroquinolines and indoles.[16] The most efficient catalyst was the montmorillonite KSF[17a] utilized under solventless conditions (Fig. 3.7).

A typical experimental procedure is as follows: the selected aromatic amine (10 mmol), phenylacetylene (1.0 g, 10 mmol) and montmorillonite KSF (1 g) were introduced in a round bottomed flask equipped with magnetic stirrer and a reflux condenser. The reaction mixture was heated at 140 °C for 5 hours and then cooled to room temperature. After addition of diethyl ether (50 ml), the catalyst was filtered and washed with diethyl ether (100 ml); the solvent was distilled off and the crude mixture was chromatographed on a silica gel column with hexane/ethyl acetate: 95/5 as eluant to give the products.

Table 3.3 shows the synthetic results obtained with different aromatic amines. Complete *ortho*-regioselective control was observed in all cases. The reaction with some classical Lewis acids such as $AlCl_3$, $SnCl_4$ and $TiCl_4$ under similar conditions afforded mixtures of tar materials accompanied by some unreacted starting anilines. *Ortho*-substituted anilines show lower reactivity with respect to the corresponding *para*-isomers, probably due to some steric hindrance troubling the interaction with the catalyst surface (Table 3.3, entries a, b, d and e). Finally, the substitution on the nitrogen atom has an important effect in the present reaction; in fact, whereas aniline gave the product **12f** with 65% yield, *N*-methylaniline **1g** afforded compound **12g** in 42% yield (entries f–g) and *N*,*N*-diethylaniline was recovered completely unchanged.

Montmorillonites are among the most efficient lamellar solid acids useful for fine chemicals preparation.[18] Their activity is mainly due to

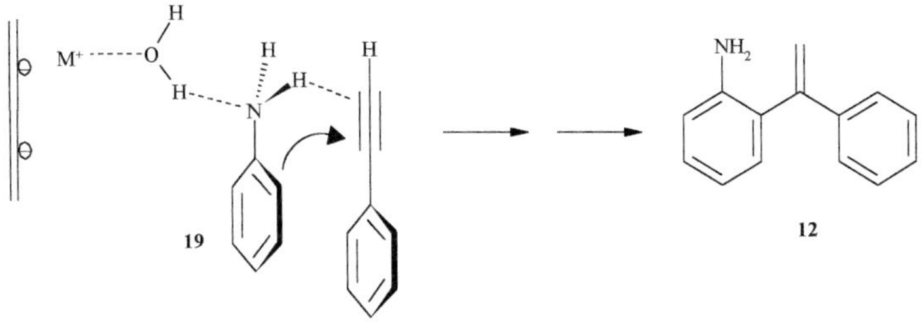

Fig. 3.8 Hypothesized mechanism of the *ortho*-regioselective alkenylation of aromatic amines.

the high surface acidity. Isomorphous substitution of Al(III) by Mg(II) or of Si(IV) by Al(III) in tetrahedral layers is responsible for both Lewis and Brönsted acidity. The Brönsted acidity is further increased by the presence of interlamellar cations required to maintain electroneutrality, which can polarize water molecules located between the negatively charged oxygens of the sheet and positively charged counter ions.[18] We suggest the mechanistic scheme depicted in Fig. 3.8 to explain reagents activation and *ortho*-regioselective control.[19]

Aromatic amines can intercalate into the layered structure of the catalyst by a variety of host–guest interactions.[20] We therefore considered that accommodation of aniline into interlayer area of the catalyst could result from H-bonding with the water polarized molecule. This interaction enhances the acidity of the N–H proton which can undergo further H-bond with phenylacetylene, producing the reactive complex **19** with a geometry suitable for a six-membered ring concerted mechanism.[21]

3.2.2 Synthesis of chromanes and *ortho-iso*pentenylphenols[22]

The alkylation of phenols with isoprene to give chromanes or *iso*pentenylphenols has received increasing attention by many researchers, but its chemo- and regioselective control is still an open problem.[23] The above result, obtained on the alkylation of aromatic substrates with phenylacetylene in the presence of solid acids, prompted us to reinvestigate the reaction of phenols with isoprene under zeolite catalysis. To this end *para*-methoxyphenol was reacted with isoprene

Table 3.3 Synthesis of various ortho-aminodiarylethylenes 12.

Entry	R	R¹	R²	Product	Yield (%)	Selectivity (%)
a	4-CH$_3$	H	H	12a	87	97
b	4-OCH$_3$	H	H	12b	93	95
c	4-Cl	H	H	12c	62	94
d	2-CH$_3$	H	H	12d	65	91
e	2-OCH$_3$	H	H	12e	50	95
f	H	H	H	12f	65	94
g	H	CH$_3$	H	12g	42	96

Table 3.4 Synthesis of variously substituted chromanes 14 and isopentenylphenols 15.

Entry	R¹	R²	R³	R⁴	14 Yield (%)	Selectivity (%)	15 Yield (%)	Selectivity (%)
a	H	H	OCH$_3$	H	65	85	45	80
b	H	H	(CH=CH)$_2$		75	85	45	85
c	H	O-CH$_2$-O		H	65	90		
d	H	H	H	H	80	90	35	75
e	H	H	OH	H	50	75		
f	CH$_3$	CH$_3$	OH	CH$_3$	65	85	53	95
g	H	OCH$_3$	OCH$_3$	OCH$_3$			40	95

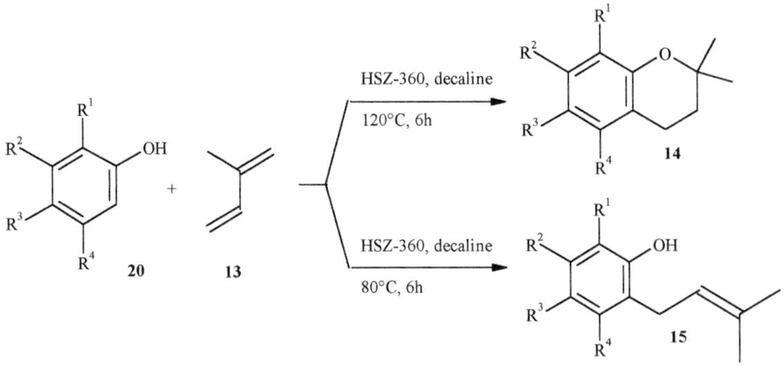

Fig. 3.9 Synthesis of chromanes and *ortho-iso*pentenylphenols.

in the presence of zeolite HSZ-360 giving the corresponding chromane in 65% yield and 85% selectivity.

This strategy has also been applied to the chromanation of other phenols. The yields usually ranged between 40 and 85% and selectivities between 75 and 90% (Table 3.4, products **14**). Next the synthesis of *ortho-iso*pentenylphenols **15** was performed by conveniently modifying the experimental conditions (Fig. 3.9). To this end the same reaction was carried out at 80 °C affording compounds **15** in 30–50% yield and 75–95% selectivity. In all cases the only side product detected in traces by gas-mass analysis (MW 136) was presumably a dimer of isoprene. A typical experimental procedure is as follows: the selected phenol (10 mmol), isoprene (1.0 ml, 10 mmol), zeolite HSZ-360 (1 g) and *ortho*-dichlorobenzene (or decaline) (50 ml) were introduced in a small autoclave and heated at the selected temperature for 6 hours under vigorous stirring and then cooled to room temperature. After filtration, the catalyst was washed with diethyl ether (100 ml), the solvents were distilled off under reduced pressure and the crude product was chromatographed on a silica gel column with hexane ethyl acetate mixtures as eluant to give the products.

Of particular interest is the synthesis of compound **15g** (Table 3.4, entry g) (40% yield, 95% selectivity), a natural product isolated from *Piper clarkii*, which was recently recognized as a powerful anticancer agent, and until now, has been synthesized in a multistep process.[24] Furthermore, in the reaction with trimethylhydroquinone we observed

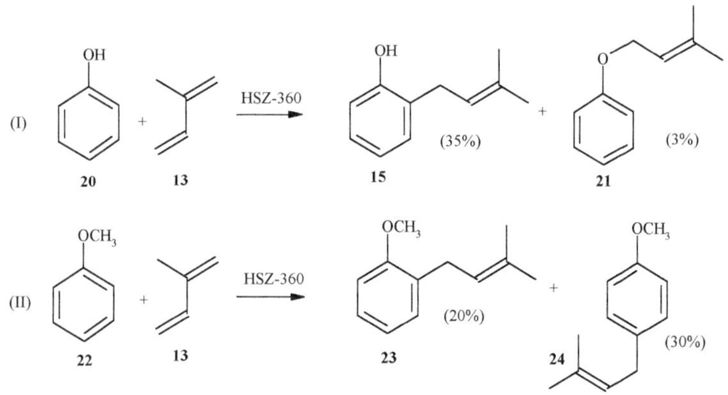

Fig. 3.10 Compared regioselectivities of phenol and anisole reactions with isoprene.

the competitive formation of the more important 2-(3-methyl-but-2-enyl)-3,5,6-trimethyl-1,4-benzoquinone which has been obtained as the only product in 53% yield by bubbling oxygen through the flask at the end of the reaction.

As also seen in Table 3.4, the process shows complete *ortho*-regioselective control regardless of products **14** or **15** obtained. Synthetic results confirm that the phenol hydroxy group plays a crucial role in the *ortho*-regioselective control. Indeed, the phenol reacted with isoprene at 80 °C for 6 h in the presence of zeolite HSZ-360 gives the *ortho*-*iso*pentenylphenol in 35% yield accompanied by a small amount of the phenyl-*iso*pentenylether **21** (3%) (Fig. 3.10). On the contrary, the same reaction with anisole **22** leads to a mixture of *ortho*- and *para*-*iso*pentenylanisoles **23** and **24** in 20% and 30% yield respectively (Fig. 3.10). Moreover, both compounds **15** and **21** afforded the chromane **14** as the sole product in 80% yield by heating the reaction mixture at 120 °C. Finally, by stirring at 80 °C a slurry of the ether **21** and zeolite HSZ-360 in *ortho*-dichlorobenzene (or decaline), *ortho*-*iso*pentenylphenol was obtained in 87% yield accompanied by a small amount of phenol (4% yield). The pathway shown in Fig. 3.11 accounts for all data observed.

Protonation of isoprene by the strongly acid zeolite HSZ-360 affords the *iso*pentenyl cation **25**, which can be envisaged as a crucial

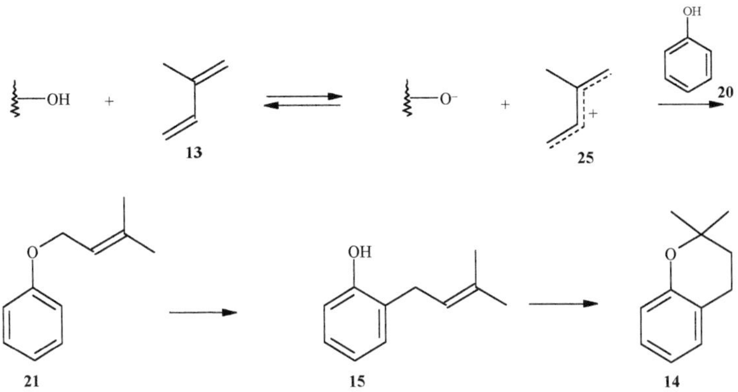

Fig. 3.11 Hypothesized mechanism of the chromane formation.

intermediate and reacts at the oxygen as the more nucleophilic site of the phenol, leading to the ether **21**. Subsequent *ortho*-regioselective [1,3] Claisen rearrangement yields compound **15** which undergoes acid-promoted cyclization to give finally chromane **14**.

This reaction shows complete regioselectivity with respect to both the aromatic substrate and isoprene; in fact the C–C bond formation exclusively involves the *ortho*-carbon of the phenol ring and the terminal 4-carbon of isoprene (see Fig. 3.11). Under homogeneous catalysis, the regioselectivity is only achieved by using sophisticated catalytic systems such as conveniently prepared acid–base combinations[10b] or transition metal complexes,[25] whereas the simple acid catalysis is reported to result in less selective reactions.[26]

3.2.3 Synthesis of chromenes[27]

In light of these results it was thought that enynes **16** would offer the potential for bisalkylation of phenol substrates leading to the construction of 2H-1-benzopyrans **17** as depicted in the comparative general scheme reported in Fig. 3.12 (Route B).

Previous studies on the chemical behaviour of α-alkynols **26** in the presence of zeolite catalysts have shown a different and completely selective process depending on the nature of the substrates **26**. Thus, heating a slurry of zeolite HSZ-360 and compounds **26** at 130 °C in

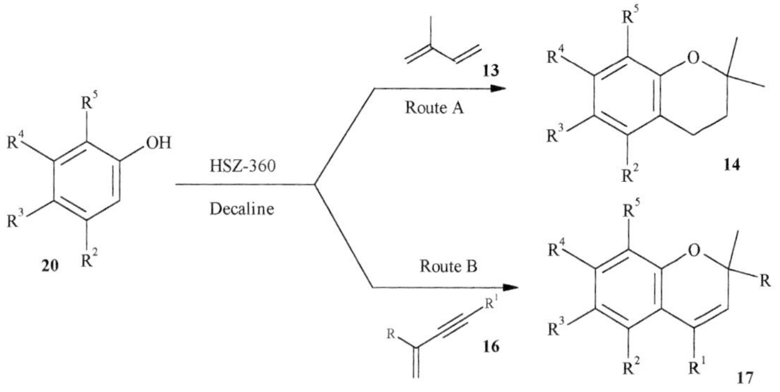

Fig. 3.12 Reaction of phenols with conjugated dienes and enynes.

chlorobenzene led to α,β-unsaturated ketones **27** or conjugated enynes **16** depending on R and R' being alkyl or aryl groups[28] (Fig. 3.13).

On the basis of these results, alkynols **26** were directly utilized as precursors of enynes **16** in the zeolite-promoted synthesis of benzopyrans **17**. Treatment of *para*-methoxyphenol with 3-phenyl-1-butyn-3-ol, the precursor of 3-phenyl-3-buten-1-yne, in the presence of zeolite HSZ-360 afforded the expected benzopyran **17cx** in 62% yield and 90% selectivity.

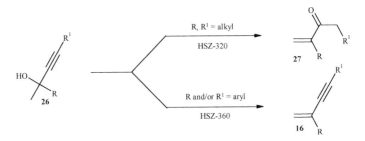

Fig. 3.13 Selective conversion of α-alkynols into α,β-unsaturated ketones or conjugated enynes.

Different α-alkynols and phenols were subjected to the following reaction conditions: the selected α-alkynol (10 mmol), the selected phenol (10 mmol) and zeolite HSZ-360 (1 g) in *ortho*-dichlorobenzene (or decaline) (10 ml) were heated at 130 °C for 6 h. After cooling to room temperature, the catalyst was removed by filtration and washed with diethyl ether (100 ml); the solvents were distilled off and the crude mixture was chromatographed on a silica gel column with hexane/ethyl acetate mixtures to give the products **17ax–gz** in high yield and selectivity. The general applicability of this reaction is shown in Table 3.5.

The key step of this process is a typical electrophilic substitution, since the presence of electron donating groups on the phenol ring results in a more facile and more efficient production of compounds **17** (Table 3.5, entries b–d). However, the process suffers from some steric hindrance. Indeed, trimethyl hydroquinone reacts with 3-phenyl-1-butyn-3-ol (**26x**) (R^1=H) affording the product **17ex** in 50% yield, whereas the reaction with 3-methyl-1-phenyl-1-butyn-3-ol (**26y**) (R^1=C_6H_5) produces compound **17ey** in 25% yield (Table 3.5, entries e and h).

The regiochemical outcome of the present reaction appears to be the same with all α-alkynols and phenols utilized since in all cases the sole products arise from interaction of the phenol oxygen with the olefinic double bond and the *ortho*-carbon of the phenol ring with the acetylenic framework of the enyne intermediate. The process shows also complete *ortho*-regiospecificity with respect to the phenol ring (Table 3.5 entry a).

Two possible modes can be envisaged by which the enyne might react with the phenol: a direct attack of the carbon carbon triple bond on the *ortho* position of the aromatic ring (Fig. 3.14 Route A) or prior aryl propargyl ether **28** formation followed by a Claisen-like rearrangement (Fig. 3.14, Route B). However, we argue against the last formulation because attempts to isolate the aryl propargyl ether in the reaction of 2,4,6-trimethylphenol with 3-phenyl-1-butyn-3-ol (**26x**) resulted in the complete recovering of the phenolic reagent accompanied by 3-phenyl-3-buten-1-yne (**16x**) in 92% yield.[29]

The acid zeolite HSZ-360 first promotes the dehydration of **26** to **16** which then regioselectively adds to the phenol substrate **20** to yield dienic compound **29**; further isomerization of **29** to tautomers **30** or **31** followed by an electrocyclic rearrangement affords the target compound **17** in agreement with previously reported studies.

Table 3.5 Synthesis of various 2H-1-benzopyrans 17.

Entry	R	R^1	R^2	R^3	R^4	R^5	Product	Yield (%)	Selectivity (%)
a	C_6H_5	H	H	H	H	H	17ax	40	90
b	"	"	"	OH	"	"	17bx	75	95
c	"	"	"	OCH_3	"	"	17cx	60	90
d	"	"	$(CH=CH)_2$		H	"	17dx	75	95
e	"	"	CH_3	OH	CH_3	"	17ex	50	85
f	CH_3	C_6H_5	H	OCH_3	H	CH_3	17cy	70	90
g	"	"	H	$(CH=CH)_2$		H	17fy	65	94
h	"	"	CH_3	OH	CH_3	CH_3	17ey	25	95
i	C_6H_5	CH_3	H	OCH_3	H	H	17cz	70	98
j	"	"	"	"	OCH_3	"	17gz	60	90

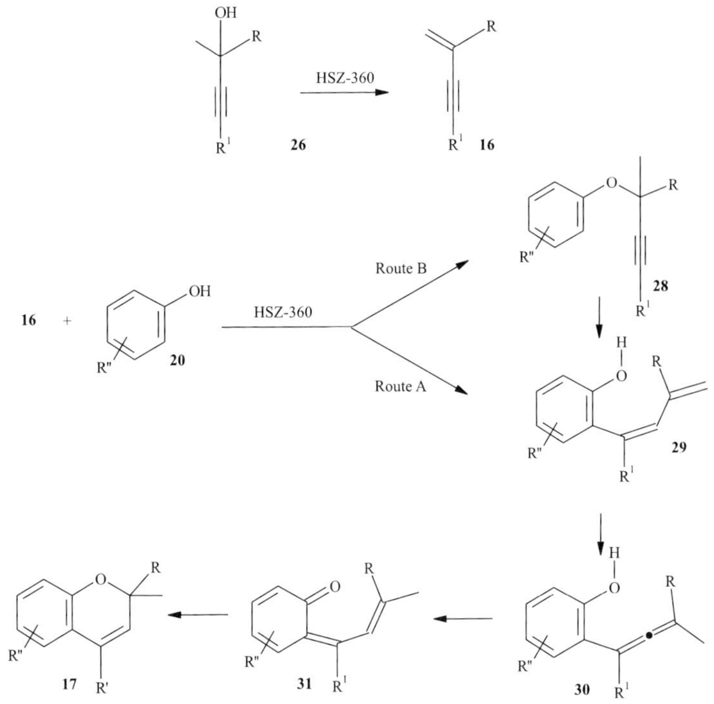

Fig. 3.14 Hypothesized mechanism of the chromenes formation.

3.3 Reaction of salicylaldehydes with malonic acid derivatives: the Knoevenagel coumarins synthesis[30]

The Knoevenagel reaction is one of the most common synthetic methods to produce coumarins. The process consists of the condensation of salicylic aldehydes with malonic acid or esters followed by decarboxylation of the coumarin-3-carboxylic acid obtained after hydrolysis. The reaction is usually catalyzed by weak bases under homogeneous conditions; amine-carboxylic acid salts and amine-$TiCl_4$ mixtures are reported to promote the process. More recently, the attention of many researchers has been focused on performing this reaction under

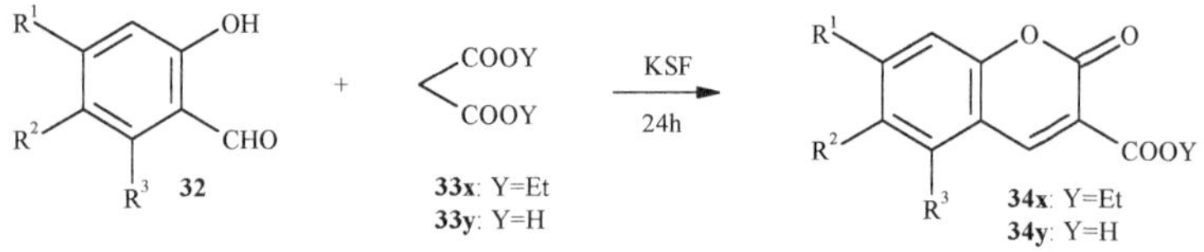

Fig. 3.15 Synthesis of coumarin-3-carboxylic acids and their ethyl esters: general scheme.

heterogeneous catalysis: in particular aluminium oxide,[31] xonotlite/*tert*-butoxide,[32] cation exchanged zeolites,[33] alcali-containing MCM-41[34] and bentonitic clay[35] have been employed. We found that montmorillonitic clays promote the Knoevenagel synthesis of coumarin derivatives in high yield and selectivity (Fig. 3.15).

In view of the importance of reducing the employment of ecologically suspected solvents, we choose to carry out the reactions in the absence of solvents or in water. Indeed there has been increasing recognition that water is an actractive medium for many organic reactions.[36] Treatment of salicylic aldehyde (**32**) with diethyl malonate (**33x**, Y=Et) in the presence of montmorillonite KSF[17a] or K10[17b] without any solvent afforded the expected coumarin-3-carboxylic acid ethyl ester.

Results suggested that the clay behaves as a ditopic catalyst containing both acid and basic sites. The basic sites, ascribable to the negative charges dispersed over entire sheets of oxygen atoms, activate the Knoevenagel condensation. Furthermore, the acid sites, mainly due to the polarized interlayer water molecules, promote the heterocycle formation by intramolecular transesterification. KSF was found to be the more efficient promoter of the reaction, probably due to its higher acidity.

It is known that the Knoevenagel reaction is strongly solvent-dependent. The first step is facilitated in solvents of high polarity and the second step, the 1,2-elimination, is inhibited by protic solvents. Thus, dipolar aprotic solvents such as dimethyl formamide are especially useful in this condensation.[37] Our results revealed that, surprisingly, the reaction is favored in aqueous medium whereas the Knoevenagel condensation is a net dehydration. Similar unexpected solvent effects in Knoevenagel reaction are reported in the literature.[34]

An additional advantage is that the aqueous medium caused the hydrolysis of the carboxylic ester functional group. In this way it becomes possible to skip the hydrolysis step before the decarboxylation reaction to coumarin. In fact, the satisfactory yield of coumarin-3-carboxylic acid (48%) was obtained by carrying out the reaction (10 mmol scale) with 3.5 ml of water. The simplest interpretation is that water brings the active methylene compound in close proximity to the negatively-charged surface of the catalyst favoring the ionization into carbanion donor both increasing the negative charge by OH dissociation and decreasing the electrostatic interaction with the interlayer counter-cations.

The fundamental role played by the catalyst surface in the present reaction was demonstrated by a comparative experiment with acid water. Thus, the reaction between salicylic aldehyde and diethyl malonate in the presence of the acid water, obtained from a stirred suspension of KSF in water after filtration of the catalyst, gave only traces of coumarin-3-carboxylic acid. Thus, the coumarin-3-carboxylic acids or their ethyl esters were accessible simply by choosing to carry out the reaction in water or without solvent.

Further the reaction could be performed by employing the malonic acid instead of its ester derivative affording coumarin-3-carboxylic acids in better yields. Thus, by reacting salicylic aldehyde and malonic acid (**33y**, Y=H) in H_2O at reflux for 24 h coumarin-3-carboxylic acid was obtained in 92% yield and 93% selectivity (Fig. 3.15). These conditions are undoubtedly milder than those previously used for malonic acid.[38]

The typical experimental procedure for the synthesis of coumarin-3-carboxylic acids (method A) is as follows: the selected salicylic aldehyde (10 mmol), malonic acid (1.6 g, 15 mmol) and KSF (1 g) in water (3.5 ml) were heated at reflux for 24 h and then cooled to room temperature. After Büchner filtration, the catalyst was washed with boiling methanol (2 × 40 ml). The solvents were distilled off and the crude product was purified, if necessary, by crystallization from ethyl acetate or methanol.

In contrast, the procedure for the preparation of coumarin-3-carboxylic acid ethyl esters (method B) is as follows: the selected salicylic aldehyde (10 mmol), diethyl malonate (2.4 g, 2.4 ml, 15 mmol) and KSF (1 g) were heated under stirring at 160 °C for 24 h. After cooling to room temperature, methanol was added (50 ml) and heated for 5 min. The catalyst was removed by filtration, washed with methanol (20 ml) and the organic phase, concentrated under vacuum, was

Table 3.6 Synthesis of various coumarins 34.

Entry	R¹	R²	R³	Y	Method	Yield (%)	Selectivity (%)
a	H	H	H	H	A	92	93
b	H	OCH₃	H	H	A	86	94
c	H	(CH=CH)₂		H	A	50	92
d	OCH₃	OCH₃	H	H	A	40	85
e	H	H	H	Et	B	44	89
f	H	(CH=CH)₂		Et	B	44	90
g	OH	H	H	Et	B	46	88

chromatographed on silica gel using mixture of hexane/ethyl acetate as eluant. Some amount of the corresponding hydrolized product was observed (~ 5%).

Table 3.6 shows the general applicability of the methods. The catalyst, simply air-dried after filtration and washing with methanol, could be reused five times without significant loss of efficiency.

References

1. Sheldon, R.A. *J. Mol. Catal. A.*, **1996**, *107*, 75–83; Clark, J.H. and Macquarrie, D.J. *Chem. Soc. Rev.*, **1996**, *25*, 303–310.
2. Tanabe, K. *Appl. Catal. A.*, **1994**, *113*, 147–152; Sheldon, R.A. *Chemtech, March*, **1994**, 38–47; Thomas, J.M. and Zamaraev, K.I. *Angew. Chem., Int. Ed. Engl.*, **1994**, *33*, 308–312.
3. See for example: Tietze, L.F. and Beifuss U. *Angew. Chem., Int. Ed. Engl.*, **1993**, *32*, 131–312.
4. Vishnyakova, T.P., Golubeva, I.A. and Glebova, E.V. *Russ. Chem. Rev. (Engl. Transl.)*, **1985**, *54*, 429–449.
5. Knölker, H.-J, Braxmeier, T. and Schlechtingen, G. *Angew. Chem., Int. Ed. Eng.*, **1995**, *34*, 2497–2500.
6. Majer, P. and Randad, R.S. *J. Org. Chem.*, **1994**, *59*, 1937–1938.
7. Bigi, F., Maggi, R., Sartori, G. and Zambonin, E. *J. Chem. Soc., Chem. Commun.*, **1998**, 513–514.
8. HSZ-360 is a commercial (Tosoh Corp.) HY zeolite with SiO_2/Al_2O_3 molar ratio 13.9, pore size 7.4 Å, crystal dimension < 0.5 μm (determined in our laboratory by SEM analysis), surface area 500 ± 10 m^2 g^{-1} (determined in our laboratory by B.E.T. method), surface acidity 0.51 meq. H$^+$ g^{-1} (determined in our laboratory by temperature programmed desorption ammonia gas (NH_3–TPD)) and with the following chemical composition (wt% dry basis): SiO_2 89.0, Al_2O_3 10.9, Na_2O 0.06.
9. Werner, W. *Tetrahedron*, **1969**, *25*, 255–261; Werner, W. *Tetrahedron*, **1971**, *27*, 1755–1760.
10. a) Casiraghi, G., Casnati, G., Puglia, G., Sartori, G. and Terenghi, M.G. *Synthesis*, **1977**, 122–124; b) Bolzoni, L., Casiraghi, G., Casnati, G. and Sartori, G. *Angew. Chem., Int. Ed. Engl.*, **1978**, *17*, 684–685.
11. Slightly lower yields were obtained by performing the reaction in decaline.
12. Sartori, G., Bigi, F., Pastorio, A., Porta, C., Arienti, A., Maggi, R., Moretti, N. and Gnappi, G. *Tetrahedron Lett.*, **1995**, *36*, 9177–9180; Arienti, A.,

Bigi, F., Maggi, R., Marzi, E., Moggi, P., Rastelli, M., Sartori, G. and Tarantola, F. *Tetrahedron*, **1997**, *53*, 3795–3804.

13. For early studies in this area see: Kolka, A.J., Napolitano, J.P. and Ecke, G.G. *J. Org. Chem.*, **1956**, *21*, 711–712.
14. Acetophenone was found to be totally ineffective in the present reaction being recovered unchanged after treatment with phenol, under the described reaction conditions.
15. Amorphous silica-alumina (SiO_2/Al_2O_3 molar ratio = 15) was prepared by adding to a stirred solution of 10% aluminium nitrate in water a colloidal silica solution (30% wt SiO_2), followed by coprecipitation of SiO_2 and $Al(OH)_3$ with NH_4OH 2M up to pH about 8.2. The solid was then filtered, washed with water, dried overnight at 120 °C and finally calcined in air at 700 °C for 16 h.
16. Guo Qiang, L. and Baine, N.H. *J. Org. Chem.*, **1988**, *53*, 4218–4222.
17. a) KSF is a commercial (Fluka) montmorillonite with surface area 15 ± 10 m^2 g^{-1}, acidity 0.85 meq. H^+ g^{-1} (determined in our laboratory by temperature programmed desorption of ammonia gas (NH_3-TPD)) and with the following chemical composition (average value): SiO_2 (54.0%), Al_2O_3 (17.0%), Fe_2O_3 (5.2%), CaO (1.5%), MgO (2.5%), Na_2O (0.4%), K_2O (1.5%); b) K10 is a commercial (Fluka) montmorillonite with surface area 200 ± 10 m^2 g^{-1}, acidity 0.70 meq. H^+ g^{-1} (determined in our laboratory by temperature programmed desorption of ammonia gas (NH_3-TPD)) and with the following chemical composition (average value): SiO_2 (73.0%), Al_2O_3 (14.0%), Fe_2O_3 (2.7%), CaO (0.2%), MgO (1.1%), Na_2O (0.6%), K_2O (1.9%).
18. Balogh, M. and Laszlo, P. *Organic Chemistry using Clays*. Springer Verlag, New York, **1993**.
19. Narayanan, S. and Deshpande, K. *J. Mol. Catal. A.*, **1995**, *104*, L109–L113.
20. Davies, J.E. *J. Incl. Phenom.*, **1996**, *24*, 133–147.
21. Ecke, G.G., Napolitano, J.P., Filbey, A.H. and Kolka, A.J. *J. Org. Chem.*, **1957**, *22*, 639–642.
22. Bigi, F., Carloni, S., Maggi, R., Muchetti, C., Rastelli, M. and Sartori, G. *Synthesis*, **1998**, 301–304.
23. Schweizer, E.E. and Meeder-Nycz, D. In *Heterocyclic Compounds. Chromenes, Chromanones and Chromanes* (ed. G.P. Ellis). John Wiley and Sons, New York, **1977**, pp. 11–139.
24. Parmar, V.S., Gupta, S., Bisht, K.S., Mukherjee, S., Boll, P.M. and Errington, W. *Acta Chem. Scand.*, **1996**, *50*, 558–560.

25. Hosokawa, T., Miyagi, S., Murahashi, S.-I. and Sonoda, A. *J. Chem. Soc., Chem. Commun.*, **1978**, 687–688.
26. Molyneux, R.J. and Jurd, L. *Tetrahedron*, **1970**, *26*, 4743–4751.
27. Bigi, F., Carloni, S., Maggi, R., Muchetti, C. and Sartori, G. *J. Org. Chem.*, **1997**, *62*, 7024–7027.
28. Sartori, G., Pastorio, A., Maggi, R. and Bigi, F. *Tetrahedron*, **1996**, *52*, 8287–8290.
29. It is important to underline that the rearrangement of aryl propargyl ethers requires high temperatures (220–240 °C) whereas the present reaction occurs at 130 °C.
30. Bigi, F.; Chesini, L.; Maggi, R. and Sartori, G. *J. Org. Chem.*, **1999**, *64*, 1033–1035.
31. Texier-Boullet, F. and Foucaud, A. *Tetrahedron Lett.*, **1982**, *23*, 4927–4928.
32. Chalais, S., Laszlo, P. and Mathy, A. *Tetrahedron Lett.*, **1985**, *26*, 4453–4454.
33. Corma A., Fornas, V., Martin-Aranda, R.H., Garcia, H. and Primo, J. *Appl. Catal.*, **1990**, *59*, 237–248.
34. Kloetstra, K.R., van Bekkum, H. *J. Chem. Soc., Chem. Commun.*, **1995**, 1005–1006.
35. Delgado, F., Tamariz, J., Zepeda, G., Landa, M., Miranda, R. and Garcia J. *Synt. Commun.*, **1995**, *25*, 753–759.
36. Li, C.J. *Chem. Rev.*, **1993**, *93*, 2023–2035.
37. Tietze, L.F. and Beifuss, U. In *Comprehensive Organic Synthesis*, (eds. B.M. Trost, I. Fleming, C.H. Heathcock). Pergamon Press, Oxford, **1991**, Vol. 2, chap. 1.11, pp. 341–349.
38. Adams, R. and Bockstahler, T.E. *J. Am. Chem. Soc.*, **1952**, *74*, 5346–5348.

4 Secondary metabolites from cells immobilized by a SiO_2 sol–gel layer

G. Carturan, R. Campostrini, and R. Dal Monte

4.1 Introduction

In chemical science, it is commonly believed that life is exclusively connected with the chemistry of carbon. But *Equisetum arvense* concentrates silicic acid of river water ($SiO_2 \approx 9$ ppm[1]) to such an extent that the dry powdered plant was once used as an abrasive by craftsmen and the husks of *Oryza sativa* hold so much SiO_2 that they constitute a non-conventional raw material for silicon carbide production.[2]

This is only one of the peculiar features of plant chemistry. Most drugs were originally used as extracts of specific plants and mosses;[3] now, chemical synthesis provides appropriate methods for obtaining those drugs at an acceptable cost for mass utilization. This synthetic route may not be feasible for metabolites with complex molecular structures or conformations, such as some *Catharanthus* alkaloids and *Taxus* diterpenes; in these cases, therapeutic exploitation is limited by the cost of extraction and purification. Advances in biotechnology may enable a general application of these drugs as: (1) *in vitro* cultivation of undifferentiated cells[4] provides impressive cell masses in place of the destruction of an unrenewable plant heritage (1 kg of taxol is extracted from 3×10^3 centuries-old trees[5]); and (2) cell immobilization on a solid support allows overall simplification of bioreactor running with increased productivity.[6]

We coupled points 1 and 2 in the case of a number of plant cells, including *Coronilla vaginalis*,[7] *Catharanthus roseus*,[8] *Taxus bacata* and *Haplophyllum patavinum*. The immobilization procedure was performed according to the Biosil® method,[9] which allows the deposition of a porous SiO_2 layer directly on the surface of plant cells supported on various scaffolding materials. This constitutes a definite contribution to 'green chemistry' for valuable drug production and large-scale utilization.

4.2 The biosil method

Si–O bond energy $= 452$ kJ mol^{-1} indicates the definite thermodynamic stability of SiO_2, which behaves as an inert material in the mild experimental conditions compatible with living cells. The structural organization of SiO_2 involves a lattice construction by oxygens bridging only two Si atoms of different tetrahedra, resulting in open tridimensional networks.

In principle, the reduction of the bridging oxygen population, attained by the simple hydrolysis reaction: Si–O–Si + H_2O → 2 Si–OH, affords a definite increase in voids in solid SiO_2. It is not conceivable to hydrolyze bulk SiO_2, but it may be possible to control the condensation of Si–OH groups of silica gels obtained by the sol–gel process.[10] According to a recent definition,[11] the sol-gel method refers to 'a broad class of processes in which a solid phase is formed through gelation of a colloidal suspension (sol)'. In the case of SiO_2 gel, Si-alkoxides are hydrolyzed affording Si–OH groups which condense, according to the following reactions:

$$-\overset{|}{\underset{|}{Si}} - OH + HO - \overset{|}{\underset{|}{Si}} - \rightarrow -\overset{|}{\underset{|}{Si}} - O - \overset{|}{\underset{|}{Si}} - + H_2O \quad (1)$$

$$-\overset{|}{\underset{|}{Si}} - OH + RO - \overset{|}{\underset{|}{Si}} - \rightarrow -\overset{|}{\underset{|}{Si}} - O - \overset{|}{\underset{|}{Si}} - + ROH \quad (2)$$

The process is crucial for the achievement of the gel state, corresponding to the immobilization of an important solvent volume in the Si–O–Si network.

Control of reactions (1) or (2) has been described in a number of cases, resulting in SiO_2 xerogels of varying specific surface area and

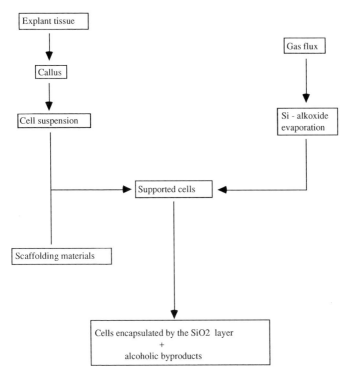

Fig. 4.1 The Biosil method.

porosity.[12,13] A useful parameter to reduce the bridging oxygen population consists of the use of Si-alkoxide holding Si–CH$_3$ or Si–H groups, which are much less suitable for hydrolysis than Si–OR. Xerogels prepared from mixtures of Si(OR)$_4$ and X$_n$Si(OR)$_{4-n}$ (X = H, CH$_3$; n = 1, 2) are known as organic-modified-silica (ORMOSIL)[14,15] and display features intermediate between SiO$_2$ and silicon rubber;[16] this fact justifies the current terminology 'hybrid SiO$_2$' to describe these products. Hybrid SiO$_2$ prepared by the sol–gel process is a valuable candidate for encapsulation of living cells: the material is stable, strong, porous and easily manufactured in thin layers. Unfortunately, SiO$_2$ gels may be obtained in mild conditions starting from solutions of alkoxide precursors which are reacted with H$_2$O to produce Si–OH and alcohol—severely toxic for any living cells.

This problem is overcome by the Biosil® method,[9] which exploits the observation reported some years ago that gaseous Si(OEt)$_4$ reacts with steam at low pressure, in the time interval of microseconds, to produce Si–OH and primary condensation products holding Si–O–Si groups.[17] As an extension, Si(OEt)$_4$ fluxing in air may react with cell surface moisture, with the formation of a silica layer directly on the cell surface (Fig. 4.1), whereas ethanol is immediately removed by the air flux.

Assessment of the experimental set-up, inherent in the consecutive steps of Fig. 4.1, leads to a primary division between: (1) laboratory work for undifferentiated cell cultures, and (2) immobilization of cells on supports suitable for continuous production in a bioreactor—avoidance of microbial contamination being the common parameter.

1. **Cell cultivation**: the preparation of undifferentiated cell cultures is a well-established biological technique.[18,19] In the nutrient medium, the explant tissue bears a callus composed of proliferating, undifferentiated cells; disaggregation of the callus produces a suspension culture which is subcultured, affording a maximum concentration of single cells or small aggregates, corresponding to a plateau in the trend of mass increase versus time. Since the rate of cell mass growth is inversely proportional to secondary metabolite productivity, we currently adopt a two-stage cultivation, according to a published method.[19] Figures 4.2 and 4.3 show examples of well-grown undifferentiated masses before disaggregation; Fig. 4.4 shows the kinetics of *Catharanthus roseus* subcultivation in suspension.

With the aim of setting up a production plant, the specific productivity of suspended cells must be tested and monitored over time. These analyses were performed on the medium, so that specific productivity refers to secondary metabolites released by the cells, in agreement with the production scheme involving recovery of products from the medium. It is generally accepted that plants accumulate secondary metabolites as defensive compounds inside the cell membrane. These compounds may or may not be released into the medium. According to Fig. 4.1, various cell lines from the explant tissues are obtained and selected on the basis of their specific productivity, measured some days after stabilization of the cell subculture at the plateau level. This demanding work is still in progress: we have stable cell lines of the studied species with valuable productivities. However, we have never achieved the production levels reported in the literature for *Catharanthus roseus*.[20]

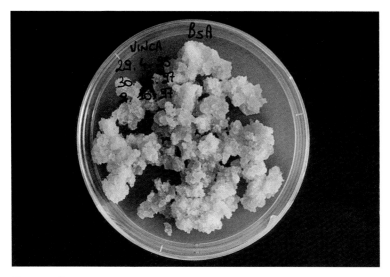

Fig. 4.2 *Catharanthus Roseus* cells obtained from stalks. Original explants were cultivated from April 29 1995, the actual masses of April 9 1998 refer to samples cultivated since October 9 1997.

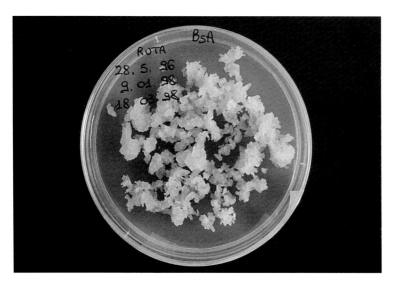

Fig. 4.3 *Haplophyllum patavinum* callus obtained from leaves. Original explants were cultivated from May 28 1996, the actual masses of April 10 1998 refer to samples cultivated since March 18 1998.

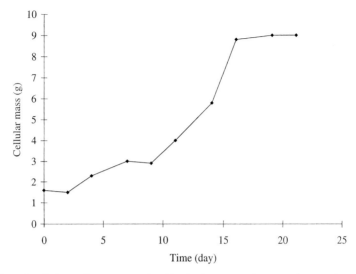

Fig. 4.4 *Catharanthus roseus* subcultivation in suspension: experiment refers to a 100 cm³ suspension of wet cells.

2. **Cells immobilization by SiO_2**: cell suspensions, with a typical cell concentration of 0.45 g of dry weight per 100 cm³ of suspension, are passed through scaffolding materials of various geometries. Materials such as glass or polyester fiber textiles, polyester sponge-like rods or cylinders, and rock wool materials can be used as scaffolds. Important common features are: (a) the dimensions of voids, which should be one order of magnitude higher than the cell diameter, (b) chemical stability, to avoid toxic interferences with living cells, and (c) mechanical strengths, for maintenance of geometrical shape in the reactor and to bear stresses due to air and medium circulation during production. These scaffolds allow the adhesion of cells so that the transition from suspended cells to supported ones is virtually quantitative. Expedients, such as using cell cultures at the growing plateau and leaving the culture medium in contact with solid supports for prolonged time, favor cell deposition and adhesion. A typical example of the cell population of *Catharanthus roseus* on polyester fibers is shown in Fig. 4.5.

Drained material is invested by an air flux saturated with silicon alkoxides, which react with membrane-sorbed H_2O to build up a sol–gel

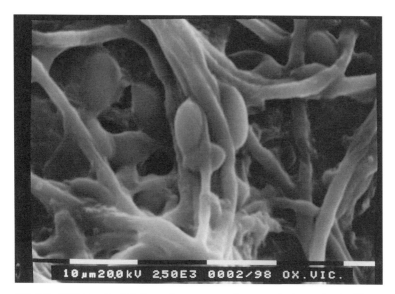

Fig. 4.5 Micrograph of *Catharanthus roseus* cells supported on polyester fibers.

SiO_2 layer. The process corresponds to a gas-to-liquid-to solid transformation *in situ*, with the formation of an inorganic material in mild conditions compatible with cell life. This type of process is also required for medical applications,[21,22] since the nanostructural features of SiO_2 (mechanical strength, tailored, porosity, chemical stability) are achieved by a process compatible with the constraints of living biological systems. The process may be considered as the inorganic counterpart of previously reported organic materials used for tissue engineering.[23,24]

The selection of alkoxides and relevant concentrations are important parameters to obtain: (1) regular SiO_2 layers, (2) pore size distributions assuring free exchange of nutrients and chemicals between cell and medium, and (3) tensile strength to immobilize the cell load to the supporting scaffolds. The basic aspects suggesting the appropriate selection of Si-alkoxides have already been published and discussed.[7,8,25] The main operative conclusions are:

- $Si(OR)_4$ must be the major component of gaseous alkoxides, since it assures the formation of a tridimensional network composed of Si–O–Si groups;

Table 4.1 Analysis of alkoxide vapour mixtures.

	$Si(Oet)_4$ / $CH_3SiH(OEt)_2$ (weight ratio)	$Si(OEt)_{4(gas)}$ (%)[a]	Total $Si_{(gas)}$ (%)[b]
1	100 / 0	100	43
2	80 / 0	25.3	97
3	70 / 30	1.6	97
4	60 / 40	1.5	97
5	0 / 100	0	100

a with reference to total Si %; b from gas chromatographic peak area of 5 = 100.

- the presence of alkoxides with Si–H bonds $CH_3SiH(OR)_2$ or $HSi(OR)_3$) favors the formation of a close SiO_2 layer. After formation of the hybrid SiO_2 layer, Si–H groups are slowly hydrolyzed in contact with the medium and transformed into Si–OH, which may condense according to reaction (1);
- saturation of the air flux with alkoxide mixtures affords a gas phase concentration which varies with time owing to the different volatility of single alkoxides.

As regards the last statement, we report here a typical experiment used to determine the actual gas phase composition of $Si(OEt)_4/CH_3SiH(OEt)_2$ mixtures. The original solution is kept at 80 °C and invested by a bubbling flux of argon = 100 ml min^{-1} in a saturator flask of 150 cm^3. The exit gas flux is analyzed by a gas chromatographic mass instrument, according to reported procedures.[26] Results (Table 4.1) are obtained after 5 min fluxing, and indicate that $CH_3SiH(OEt)_2$ (boiling point 94–5 °C) is the major component in the starting flux, which is progressively enriched by $Si(OEt)_4$ (boiling point 160 °C) with a concomitant decrease in the instantaneous concentration of total Si species. These observations are important, because the operative conditions, suitable to control the SiO_2 layer formation, can be adjusted or foreseen from data collected in a similar way to those in Table 4.1, for various alkoxide mixtures.

As previously reported,[7,22] the thickness of the deposited layer may be determined by the time of exposure of the supported cells to the reactive alkoxide flux.

Figure 4.6 shows some *Catharanthus roseus* cells coated by the SiO_2 layer. Besides the morphological evidence of porous SiO_2 coating the

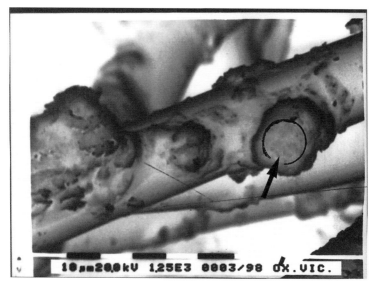

Fig. 4.6 Micrograph of *Catharanthus roseus* cells after entrapment by a hybrid sol-gel SiO$_2$ layer.

cell surface, Si elemental analysis in Fig. 4.6 (arrow) highlights the presence of silicon (Fig. 4.7).

4.3 Productivity of SiO$_2$-encapsulated cells

The biological mechanism producing secondary metabolites is mostly unknown, but it certainly changes from species to species. Natural productivity is affected by several factors such as climate, habitat and vegetative cycle. As already mentioned, most plant accumulate secondary metabolites for defense or produce them under stress. These features are unpredictable production parameters which cannot be applied, for example, to a bioreactor holding immobilized cells; therefore, preventive selection of productive cell lines appears to be the easiest and most inexpensive approach. Moreover, undifferentiated cell cultures having good production of metabolites which remain confined in the cells reduce the advantages of immobilization, productivity being determined by drug concentration in the medium.

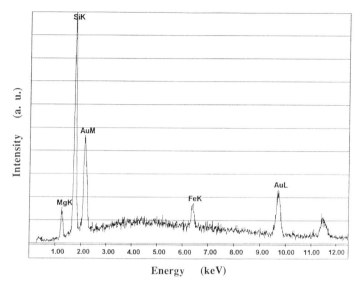

Fig. 4.7 Si elemental analysis by XRF-EDMA on the 50 μm^2 surface (see arrow in Fig. 4.6). Mg and Fe are present in the original medium.

Preventive selection of suitable cell lines is an important parameter to quantify the revenue versus cost balance.

Experiments were performed on *Coronilla vaginalis* productivity of coumarins,[7] *Catharanthus roseus* alkaloids,[8] *Taxus bacata* taxanes, and *Ruta p.* secondary metabolites, exhaustive results being collected only in the case of *Catharantus roseus*. Productivity obtained with the bioreactor of Fig. 4.8 refers to 200 g of wet cells encapsulated on a rock-wool scaffold by a hybrid SiO_2 layer 0.3 μm thick, obtained from deposition of gaseous $Si(OEt)_4/CH_3SiH(OEt)_2 = 80/20$ (nominal weight ratio of the liquid mixture). The total alkaloid production versus time follows a S-type profile, like that of *Coronilla vaginalis* productivity[7] and of non-encapsulated cells.[27] The maximum concentration, corresponding to the production plateau, is reached after 26 ± 2 days and is 0.23–0.26 g of total alkaloids per g of dry cell mass. This result appears exceptional in comparison with the production from our non-encapsulated cells, which was 20–23 times lower.[8] This beneficial effect of encapsulation was expected, in agreement with the generally

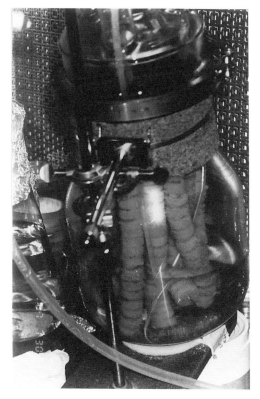

Fig. 4.8 Experimental 10 dm³ bioreactor used for *Catharanthus roseus* metabolite production by encapsulated cells.

observed increase in output of bioreactors composed of immobilized cells,[28,29] as found by us for *Coronilla vaginalis*[7] and for glucose fermentation in *Saccaromices cerevisiae* SiO_2-immobilized cells.[25] According to these production kinetics, the medium is substituted every 24–26 days and the process set-up is duly adjusted.

Taxus bacata immobilized cells displays a productivity of taxanes of 0.95 mg per g of dry cell mass per day; the maximum concentration of 3.8 mg dm⁻³ was reached in four days in our 1.2 dm³ experimental reactor holding 18.6 g of wet cells ($\approx$ 0.93 g of dry mass). This rapid saturation may be attributed to the poisoning effect of taxanes on cellular metabolism[30] or to the low solubility of these chemicals in the medium. These preliminary data are awaiting comparison with results reported in a recent excellent review.[31]

The cases of *Catharanthus roseus* and *Taxus bacata* may be considered as two opposite models for process set-up, since the former affords secondary metabolites in high concentration at a frequency of 24 ± 2 days, and the latter bears products in low concentrations at a 4-day frequency. Possible production processes (Fig. 4.9A and B) are presented below. Extraction and separation of *Catharantus roseus*

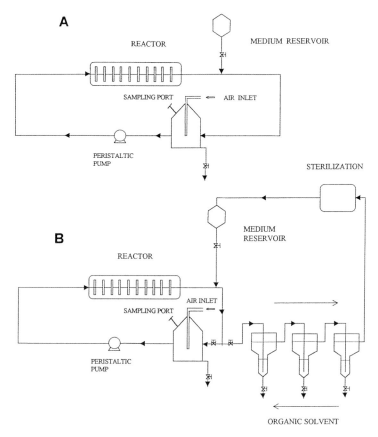

Fig. 4.9 Production process for entrapped cells of *Catharanthus roseus* (A) and *Taxus bacata* (B).

alkaloids from the medium are well documented and the discontinuous process of Fig. 4.9A, with recovery of alkaloids and discharge of the medium, is reasonable owing to the high alkaloid concentration and low frequency of medium substitution. The scheme (Fig. 4.9B) proposed and profitably used for taxanes, is explained by opposite considerations. In particular, the use of chlorocarbons as extraction solvents is very favorable, owing to the very high extraction selectivity versus taxanes and the low distribution coefficient versus sucrose and most additives of the medium.

The study of *Ruta p.* secondary metabolites is still in progress. In the research-scale conditions (24 g of wet cells in a 1.3 dm^3 reactor) valuable products are lignans, recognized by HPLC separation and mass-spectrometry analysis. The peculiarity of these species is the exploitation of the high vapour tension, which allows the separation of an important fraction of lignans upon fractional condensation from the medium vapours obtained on heating at 48–55 °C.

4.4 Subculturing versus mortality rate

Rational management of any production plant demands exact knowledge of the mass balance. In the case of metabolite production by immobilized cells, the mass balance between products and reagent chemicals cannot be performed, since the stoichiometric reaction sequence is often unknown; however, the balance between cell mass production and mortality rate is essential to determine the productivity schedule of a specific metabolite, and space, time and personnel ratios between the biological mass production unit and that of the production plant. This requirement involves determination of cell mass kinetic growth, similar to the trend shown in Fig. 4.4, and the mortality rate of the SiO_2-encapsulated cell load. This latter experiment was carried out for *Catharanthus roseus*, monitoring the sucrose consumption of 200 g of wet cells in a 10 dm^3 reactor. The sucrose consumption kinetics follow a zeroth-order law (Fig. 4.10) up to 80 days, resulting in a specific rate constant = 240 ± 3 mg $l^{-1}day^{-1}$ per gram of dry weight cells.[8] Deviation from linearity above 80 days was considered an index of aging + mortality of the original cell load, whose reproduction was hampered by SiO_2 entrapment. With this assumption, the difference, Δ, between expected and experimental consumption may substantiate the

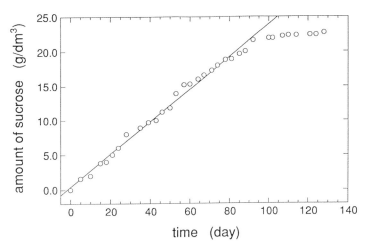

Fig. 4.10 Sucrose consumption of 200 g of wet cells of *Catharanthus roseus* after encapsulation (total volume = 10 dm^3).

advance of cell mortality. As shown in Fig. 4.11, the acceptable linearity of ln∆ versus time gives a specific rate constant = 0.042 ± 0.002 day^{-1} per gram of dry weight, corresponding to about 4% of daily mortality after 77 ± 5 days from immobilization, and to $t_{1/2} = 16.3$ days. These results allow us to calculate the life expectation of the immobilized cells according to the equation: life expectation = $75 \pm 5 + 3\, t_{1/2} = 125 \pm 5$ days, which agrees with the seasonal life of *Catharanthus roseus* species. The cellular mass balance resulting from these data is as follows:

productivity: 0.4 g of valuable metabolites (vincristine + vinblastine) per 10 g of dry cells in 100 days;
reactor: 10 g of dry cells (200 g of wet cells), volume 10 dm^3;
immobilization time: 3 days;
cell aging: 6 days;
mass balance: 10 g of wet cells for subculture for about 100 days.

Experiments to quantify the productivity and mass balance of our *Taxus bacata* cell line are still in progress. This species displays only 20% of cell weight increase during 21-day subculturing.

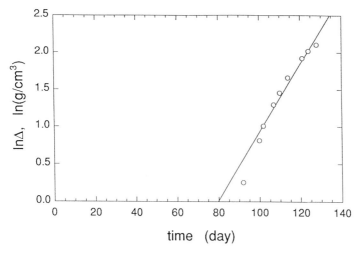

Fig. 4.11 Ln (Δ = mortality) versus time, after deviation from linearity of sucrose consumption.

4.5 Rough revenue estimate and conclusions

The mass balance given above and other reported data allow us to estimate the revenue of a bioreactor composed of SiO_2-entrapped plant cells. It is a generally accepted parameter that the revenue of a bioreactor of 0.15 $ per litre per day is an economically feasible criterion, independent of the specific product.[19] Considering a bulk market price of 100 $ g for vincristine and vinblastine (40% of our total alkaloid production), the results of our laboratory-scale reactor lead to a revenue of 0.04 $ that is 3.8 times smaller than the above criterion.

This valuation, although unsatisfactory, indicates that selecting more productive cell lines is the correct approach to increase revenue to 0.15 $ per liter per day. Indeed, the cell mass productivity of *Catharantus roseus* undifferentiated cultures is confidently expected to yield one cell line with productivity similar to the values reported 15 years ago by J.P. Kutney *et al.*[20]

The revenue for productivity of the bioreactor using *Taxus bacata* immobilized cells cannot at present be estimated, since our production data refer to a small apparatus (1.2 dm^3) holding only 18.6 g of wet cellular mass, producing 0.95 mg per g of dry cell mass per day of taxanes. In conclusion, hybrid sol-gel SiO_2 entrapment of plant cells according to the Biosil method appears to be a valuable approach for immobilization of large cell masses. The maintenance of viability and study of the mass balance between cell subculturing and cell mortality may be significant parameters for the extension of biological 'green chemistry' to the production of pharmaceutical products.

References

1. Riley, J.P. and Chester, R. (1971). *Introduction to marine chemistry*. Academic Press, New York.
2. Mishra, P.K., Nayak, B.B., and Mohanty, B. (1995). Synthesis of silicon carbide from rich husk in a packed bed arc reactor. *Journal of American Ceramic Society*, **78**, 2381–4.
3. Abelson, P.H. (1990). Medicine from plants. *Science*, **247**, 513.
4. Kutney, J.P. (1993). Plant cell culture combined with chemistry: a powerful route to complex natural products. *Accounts of Chemical Research*, **23**, 559–66.
5. Nicolau, K.C., Dai, W.M., and Guy, R.K. (1994). Chemistry and biology of taxol. *Angewandte Chemie International Edition English*, **33**, 15–44.
6. Vandrey, C. (1995). Why immobilize? *Immobilized cells: basic and applications*, (ed. R.H. Wijffels, R.M. Buitelaar, C. Bucke, and J. Tremper), pp. 3–16, Elsevier, Amsterdam.
7. Campostrini, R., Carturan, G., Caniato, R., Piovan, A., Filippini, R., Innocenti, G., and Cappelletti, E.M. (1996). Immobilization of plant cells by hybrid sol–gel materials. *Journal of Sol-Gel Science and Technology*, **7**, 87–97.
8. Carturan, G., Dal Monte, R., Pressi, G., Secondlin, S., and Verza, P. (1998). Production of valuable drugs from plant cells immobilized by hybrid sol-gel SiO_2. *Journal of Sol-Gel Science and Technology*, **13**, 273–276.
9. Cappelletti, E.M., Carturan, G., and Piovan, A., (21 November 1996). Process and articles for producing secondary metabolites of viable plant cells immobilized in a porous matrix. *PCT International Publication* No 96/36703, patent Nr. 95920202.9-1222.
10. Brinker, J. and Scherer, G. (1990). *Sol-gel science*, Academic Press, San Diego.
11. Lev, O., Wu, Z., Bharathi, S., Glezer, V., Modestov, A., Gun, J.,

Rabinovich, L., and Sampath, S. (1997). Sol–gel materials in electrochemistry. *Chemistry of Materials*, **9**, 2354–75.

12. Carturan, G., Coccco, G., Schiffini, L., and Strukul, G. (1980). Role of molecular interactions during the preparation of Pd and Pt/glass hydrogenation catalysts, in determining the physical state and chemical reactivity of the metal. *Journal of Catalysis*, **65**, 359–68.

13. Carturan, G., Facchin, G., Gottardi, V., and Navazio, G. (1984). Preparation of supports for catalysis by the gel route. *Journal of Non-Crystalline Solids*, **63**, 273–81.

14. Schmidt, H., Scholze, H., and Kaiser, A. (1984). Principles of hydrolysis and condensation reaction of alkoxysilanes. *Journal of Non-Crystalline Solids*, **63**, 1–11.

15. Huang, H., Orler, B., and Wilkes, L. (1987). Structure-property behavior of new hybrid materials incorporating oligomeric species into sol-gel glasses. 3. Effect of acid content, tetraethoxysilane content, and molecular weight of poly(dimethylsiloxane). *Macromolecules*, **20**, 1322–30.

16. Dirè, S., Pagani, E., Babonneau, F., Ceccato, R., and Carturan, G. (1997). Unsupported SiO_2-based orgnaic-inorganic membranes. Part 1. Synthesis and structural characterization. *Journal of Materials Chemistry*, **7**, 67–73.

17. Campostrini, R., Carturan, G., Pelli, B., and Traldi, P. (1989). Hydrolysis and polycondensation of $Si(OEt)_4$: a mass spectrometry approach to the definition of reaction intermediates in the presence of acid or basic catalysts. *Journal of Non-Crystalline Solids*, **108**, 143–9.

18. Charlwood, B.V. and Rhodes, M.J.C. (1990). *Secondary products from plant tissue culture*, Oxford Science, Oxford.

19. Payne, G.B., Bringi, V., Prince, C.L., and Shuler, M.L. (1992). *Plant cell and tissue culture in liquid systems*. Hauser, Munich.

20. Kutney, J.P., Aweryn, B., Choi, L.S.L., Honda, T., Kolodziejczyk, P., Lewis, N.G., Sato, T., Sleigh, S.K., Stuart, K.L., Worth, B.R., Kurz, W.G.W., Chatson, K.B., and Constabel, F. (1983). Studies in plant tissue culture: the synthesis and biosynthesis of indole alkaloids. *Tetrahedron*, **39**, 3781–95.

21. Carturan, G., Muraca, M., and Dal Monte, R. (28 May 1996). A process for encapsulating viable animal cells. Application Nr. PCT/EP 96/02265.

22. Carturan, G., Campostrini, R., Vilei, R.T., Dal Monte, R., Zanusso, E., and Muraca, M. (1997). Encapsulation of viable animal cells by SiO_2: chemical advances and prospectives for hybrid bioartificial organs. In *Bioartificial Liver Support Systems*, pp. 48–55, CIC Edizioni Internazionali, Rome.

23. Hubbel, J.A. (1996). In situ material transformation in tissue engineering. *MRS Bulletin*, November, pp. 33–35.
24. Andreopoulos, F.M., Deible, C.R., Stauffer, M.T., Weber, S.G., Wagner, W.R., Beckman, E.J., and Russel, A.J. (1996). Photoscissable hydrogen synthesis via rapid photopolymerization of novel PEG-based polymers in the absence of photoinitiators. *Journal of the American Chemical Society*, **118**, 6235–40.
25. Inama, L., Dirè, S., Carturan, G., and Cavazza, A. (1993). Entrapment of viable microorganisms by SiO_2 sol-gel layers on glass surface: trapping, catalytic performances and immobilization durability of *Saccaromyces cerevisiae*. *Journal of Biotechnology*, **30**, 197–210.
26. Campostrini, R., D'Andrea, G., Carturan, G., Ceccato, R., and Sorarù, G.D. (1996). Pyrolysis study of methyl-substituted Si–H containing gels as precursors for oxycarbide glasses, by combined thermogravimetry, gas chromatographic and mass spectrometric analysis. *Journal of Materials Chemistry*, **6**, 585–94.
27. Stafford, A. and Smith, L. (1986). Effects of modification of the primary precursor level by selection and feeding on indole alkaloid accumulation in *Catharanthus roseus*. *Secondary metabolism in plant cells cultures*, pp. 250–56, University Press, Cambridge.
28. Divies, C. and Siess, M.H. (1981). Behavior of *Saccaromyces cerevisiae* cells entrapped in a polyacrylamide gel and performing alcoholic fermentation. *European Journal of Applied Microbiology and Biotechnology*, **12**, 10–5.
29. Asada, M. and Shuler, M.L. (1989). Stimulation of ajmalicine production and excretion from *Catharanthus roseus*: effects of adsorption *in situ*, elicitors, and alginate immobilization. *Applied Microbiology and Biotechnology*, **30**, 475–84.
30. Gibson, D.M., Ketchum, R.E.B., Vance, N.C., and Christen, A.A. (1993). Initiation and growth of cell lines of *Taxus brevifolia* (Pacific yew). *Plant Cell Rreproduction*, **12**, 479–86.
31. Gibson, D.M., Ketchum, R.E.B., Hirasuna, T.J., and Schuler, M.L. (1995). Potential of plant cell culture for taxane production. In *Taxol: science and applications*. (ed. M. Suffness), pp. 71–95, CRC Press, Boca Raton.

5 *The activation of hydrogen peroxide for selective, efficient wood pulp bleaching*

Terrence J. Collins, Jenny Hall, Leonard D. Vuocolo, Nadine L. Fattaleh, Ian Suckling, Colin P. Horwitz, Scott W. Gordon-Wylie, Robert W. Allison, Terry J. Fullerton, and L. James Wright

5.0 Prologue

Much of the global pulp and paper industry (P&PI) has been making significant progress in pollution reduction in recent years. This progress has been proceeding at considerable capital expense and the transformation that is resulting is a tribute to the industry. Our discussion necessarily focuses on specific and general pollutants produced in papermaking and it is cast in the mold of a green chemistry master strategy. This strategy seeks to move the elemental balance of pulp bleaching closer to what Nature employs for degrading lignin,[1] a strategy reflected in the industry's recent development of totally chlorine free (TCF) bleaching procedures. Our purpose here is to point out that a new technology is developing for papermaking that holds promise for enabling further pollution reduction in the industry in a cost-effective manner that ensures the quality of the bleached pulp product. The technology is labeled 'P_{Fe}' to denote that it is a peroxide bleaching process activated by iron catalysts. We make this presentation with the conviction that any new bleaching technology should not be forced on an industry that has recently mortgaged itself heavily for the sake of the environment. Rather, process adaptations should be left as decisions for the P&PI to make based upon the balance of environmental, economic and technical attributes of any newcomer to the technology base.

5.1 Introduction

5.1.1 Following the elemental balance of Nature in chemical processes: a significant green chemistry[2] strategy

Selectivity is of paramount importance both in chemistry and in life processes, but chemists and Nature apply fundamentally different strategies to achieve it. We chemists practice relatively simple reagent design and employ almost the entire periodic table to attain reactivity objectives. In contrast, Nature succeeds with selectivity goals by employing elaborate design based upon the limited set of elements available in each environment. This strategic difference is the genesis of an antagonism that underlies much of the pollution attributable to chemistry. We will begin this chapter by reflecting on several examples of the antagonism. Each example teaches the same lesson, one that enjoins chemists wherever possible to move the elemental balance of technology closer to that which Nature uses for the same or a related process. Stated alternatively, each example reinforces that this movement in the elemental balance of technology is an important order of business for green chemistry.

As a working definition of the term 'pollution', we consider that the environment is polluted by chemistry when, by inducing a change in a life-supporting chemistry, the practice of a chemical technology jeopardizes any life-form that is recognized to be important to us.[3] This definition does not include all life forms because we must defend ourselves from certain organisms, such as the bacteria that bring us fatal diseases. It is life that is the most precious element of our reality and the human desire to protect life is the most powerful force aligning our civilization behind green chemistry. When pollution is recognized and genuinely understood, the power of reason maintains pressure on the case for getting rid of the source, all the way to complete elimination. While, in the short term, prudence may urge for an economically feasible response, there is ample historical precedent that, in the long term, the case for environmental protection is an immutable force to which an economy must adapt.

To illustrate the adaptation process, we can see that many chemical technologies have been modified or abandoned because the presence of an element or elements caused unacceptable pollution. Often the process proceeds in slow motion. For example numerous serious cases of pollution involve lead. Recently, an interesting story concerning lead pollution was related by Josef Eisinger.[4] In the early 1690s, the official physician of the city of Ulm, Eberhard Gockel, rightly identified the practice of sweetening acid wine, a practice called 'correcting wine', as the cause of the painful and often fatal ailment, colic. The sweetening agent was obtained by dissolving litharge, PbO, in a quantity of wine that was then evaporated to give a sweet buffering extract. While Ulm and its environs were subsequently spared the colic, for several centuries over much of the rest of Europe, colic epidemics continued to follow in the footsteps of the adverse weather conditions for grape growing that induced vintners to correct their wines. Eventually, Gockel's insight came to be widely acknowledged and the great European colic epidemics finally ceased. Lead is now very widely recognized as being a harmful element to humans.[5] More recently, lead compounds have been significantly replaced in gasoline with carbon-based antiknock agents and in paints with non-lead flowing, preserving, coloring and binding agents, although not yet on a global basis in either technology area. Societies that tough it out and replace lead wherever possible in their technologies, especially those that contribute significantly to human exposure, undoubtedly improve the quality of life and the potential of their people.[5]

In the late twentieth century, extensive knowledge exists concerning elemental toxins and corrective action is usually possible. Because the toxicity is a property of the element, such compounds can never be detoxified; they can only be contained or not dispersed in the environment in the first place. Cases of polluting technologies also abound where persistent, harmful chemicals arise from the presence of a relatively non-toxic element in a particularly stable composition. The pollution may come in the direct form of a particular or a universal toxicity or it may compromise the environment in such a way that a threat to life is posed indirectly. The deleterious compounds can occur as products that have been deliberately synthesized to support a particular technology and are later found to have harmful properties. They also occur as by-products of large-scale chemical technologies that have been executed with a non-natural elemental balance. Some such cases provide examples of successful green chemistry while the others provide significant targets for green chemistry.

Despite its common use in Nature, chlorine has a conspicuous and varied history in this area of pollution, arising from mankind's extensive use of the element in large-scale global technologies over the last two centuries. Important cases of polluting primary products and polluting by-products can be found among them. The chlorofluorocarbon/stratospheric ozone depletion[6,7] predicament provides an important example of polluting primary products. As is well-known, the problem here was created by volatile chlorine-containing organics being too robust, one of the very properties that initially made the CFCs so seemingly attractive for tropospheric use as refrigerants and blowing and cleaning agents. The case also illustrates modern green chemistry in comparatively rapid action.[8] The ability of the CFCs to deliver ozone-depleting chlorine to the stratosphere is now widely recognized. The problem is being vigorously corrected in many countries by replacement of CFCs first with weaker ozone depleters and then with ozone non-depleters. The ability of each refrigerant and blowing agent to contribute to green house warming is also being factored into replacement strategies; the warming ability is being analyzed as a component of the overall warming potential of each refrigeration technology.[8]

The serious negative environmental and health issues associated with other persistent chlorinated chemicals such as the contact insecticide DDT, the PCBs, and the polychlorinated dioxins and dibenzofurans have also been significantly learned (Fig. 5.1).[9,10] Toxic PCBs[11] have been used extensively in technology in electrical transformers and capacitors, gas transmission turbines, vacuum pumps and heat transfer systems, and as hydraulic fluids, plasticizers, adhesives, fire retardants, wax extenders, dedusting agents, pesticide extenders, inks, lubricants and cutting oils. Residual PCBs from these technologies present significant disposal liabilities.[11] Polychlorinated dioxins and dibenzofurans can be produced upon incineration of organic matter in the presence of chlorine. The presence of metal ions such as copper is also required.[12] The chlorine can be in any form in the incineration mixture. These same persistent toxins arise from chlorine-based bleaching in the P&PI.[10]

The resolution process with persistent toxins is usually difficult. As Eisinger relates,[4] in 1696 when Duke Eberhard Ludwig of the duchy of

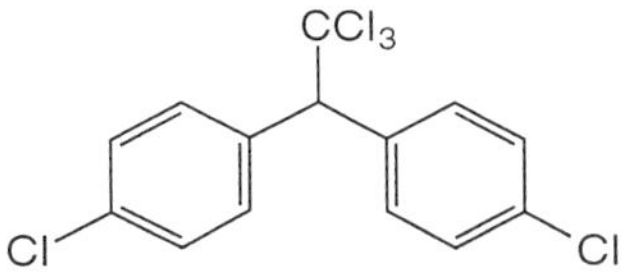

DDT
1,1,1-Trichloro-2,2-bis(p-chlorophenyl)ethane

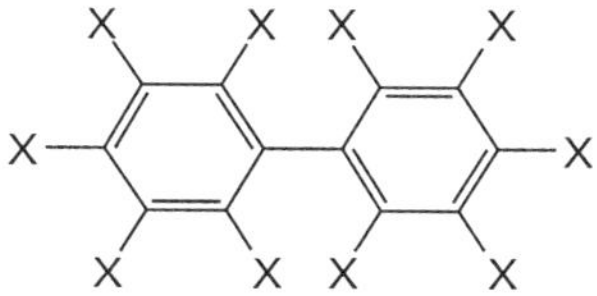

PCB
Polychlorinated Biphenyls (X = H, Cl)

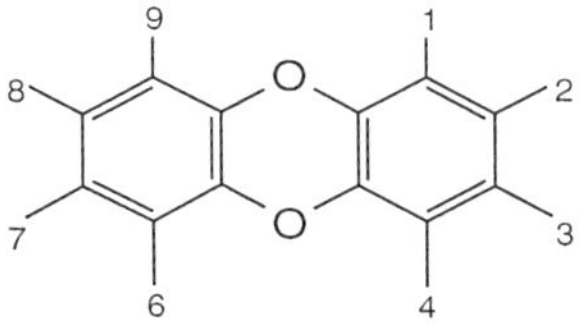

Chlorinated Dibenzodioxin
(substituents 1-4 and 5-9 = H, Cl;
2,3- and 7-8-chlorinated species are toxic)

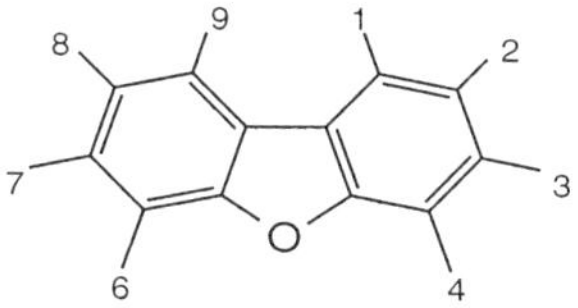

Chlorinated Dibenzodifuran
(Substituents 1-4 and 5-9 = H, Cl;
2,3- and 7-8-chlorinated species are toxic)

Fig. 5.1 Structures of persistent chlorinated pollutants.

Württemburg was told of Gockel's still unpublished findings on 'wine disease', he quickly issued an edict banning all lead-based wine additives. Violation was made punishable by death and the same punishment was extended to anyone knowingly failing to report a perpetrator. This decisive and harsh response to eradicate a polluting technology served both consumer protection and the economic security of Ulm. The economy of late seventeenth century Ulm was based not on wine production, but on the wine trade; the city's wealth was threatened when people began to blame the colic on wines being traded there. Rarely do economic factors so effectively reinforce pollution reduction goals. Instead, pollution reduction generally burdens economies, especially in the short term. In the wine disease case, the economic hardship arose primarily outside the city in the wine producing regions that marketed their products through Ulm's markets. But there is another important lesson in these early examples of chemistry. Banning wine correcting has not destroyed the wine industry. Replacing CFCs has not incapacitated economies in the areas of refrigeration and the many other fields where CFCs used to play a vital role. Eliminating lead has not eliminated the paint industry. Green chemistry can indeed work very positively for everyone.

As the new millennium approaches, chemists have a major role to play in reducing persistent toxins while minimizing the negative and maximizing the positive industrial and economic consequences. As Paul Anastas and others have eloquently articulated, green chemistry has great potential for allowing mankind to achieve healthy, sustainable development.[2,3,13–16]

5.1.2 Bleaching in the pulp and paper industry (P&PI)

While modern pulp bleaching processes are sophisticated in their engineering, they are remarkably simple in the practice of their chemistry. This simplicity presents an enticing challenge to chemists because it should be possible to improve the processes significantly through catalysis. We continue by first giving the reader a background to the history of the chemicals used in pulp bleaching while touching on the far-reaching and successful efforts of the pulp and paper industry to reduce chlorinated effluents. Then, we will describe P_{Fe} bleaching technology as a new way to enable the move of a large-scale global industry toward a natural elemental balance in its applied chemistry. The background information that follows has been extracted primarily from several excellent books on the subject of wood pulp bleaching.[17–20] Raw chemical pulp prior to bleaching consists primarily of the cell wall components of wood after most of the intracellular materials or extractives and the lignin binding agent have been removed. Chemically, this pulp is comprised mainly of colorless cellulose, which is a homopolysaccharide of D-glucose (Fig. 5.2), and colorless hemicellu-

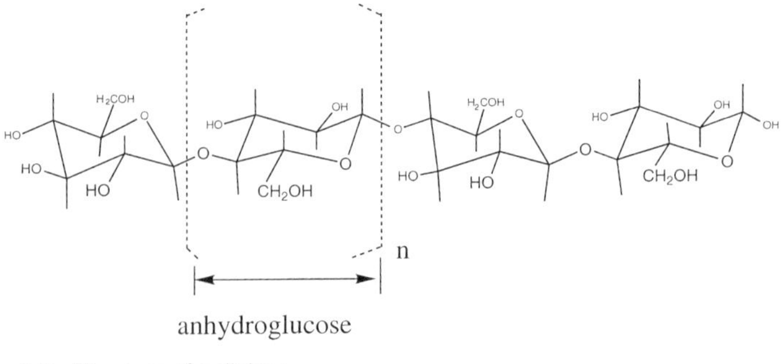

Fig. 5.2 Structure of cellulose.

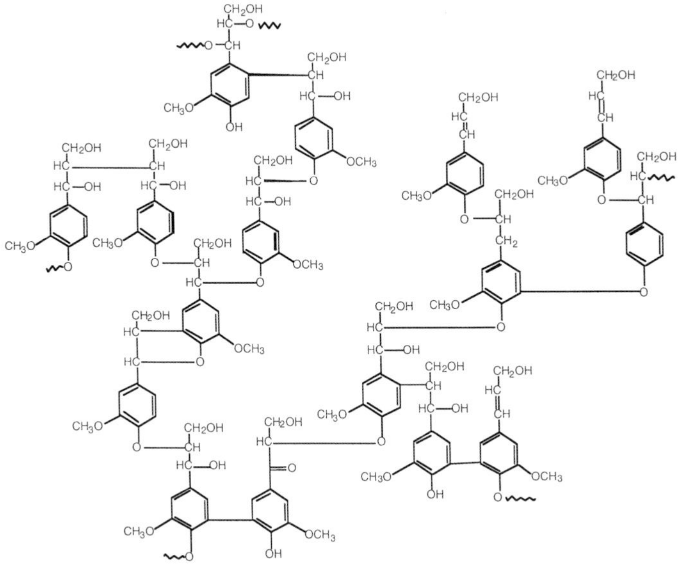

Fig. 5.3 A hypothetical depiction of a portion of a softwood lignin polymer adapted from Biermann.[40]

loses, which are heteropolysaccharides, and the source of color, residual lignin (Fig. 5.3). 'Lignin' is the name given to a collection of polymers of wood-dependent composition consisting mainly of monomers of oxidized phenylpropane that differ in the way they are oxidized and that are linked apparently in non-ordered fashions. Lignin must be removed from the cellulosic materials to produce white fiber for papermaking.

The dominant chemical technology for obtaining pulp suitable for bleaching en route to quality paper is called the kraft process. In this treatment, wood chips are impregnated with pulping liquor, a solution of NaOH and Na_2S (3 : 1 to 4 : 1 with both chemicals expressed as Na_2O equivalents), at a liquor-to-wood ratio of about 4. The mixture is then heated to 150 to 180 °C for 1–2 hours. Pulping breaks down most of the lignin, which becomes dissolved such that it can be carried off in the extracted liquor, the so-called 'black liquor'. Laden with lignin fragments, some cellulosic material and salts, the black liquor is burned in the recovery boiler of the mill to generate energy and to return the majority of the pulping salts for reuse. The energy generation and salt recovery are important components of the financial feasibility of kraft mills.

The kraft process returns a yield of 40–55% of crude pulp based upon the wood input in a form that is soft and easily fiberized. At this point, depending on the wood source and extent of the treatment, the pulp lignin content has been reduced from 18–35% to 2–6%. Oxidative bleaching follows to remove this residual lignin with the goal of obtaining bright white cellulosic fiber for quality papermaking. Here we will use the term 'bleaching' to imply both removal of residual lignin and brightening. Obtaining high brightness is the principal goal of chemical bleaching. Brightness is a term used to describe the whiteness of pulp on a scale from 0% (absolute black) to 100% (relative to a MgO standard, which has an absolute brightness of ca. 96%). Brightness is determined by the reflectance of blue light (457 mm) from the paper produced from the pulp. Removal of all the lignin leads to bright white pulp that does not yellow on aging.

From the end of the 18th century, dominant oxidation technologies for papermaking have been provided by chlorine-based oxidants. Hypochlorite bleaching (labeled H bleaching by the P&PI) was discovered at the close of the 18th century and was the standard bleaching agent for over a century. Hypochlorite was both supplemented and gradually replaced by elemental chlorine (C bleaching) in the 19th and 20th centuries. Chlorine began to be complemented by chlorine dioxide (D bleaching) in the 1940s because this relatively expensive bleaching agent is highly selective for attack at the lignin over the cellulose, all the way to complete lignin removal and high brightness. Pulp bleaching involves multiple bleaching, extraction and washing cycles. In the earlier cycles, when more lignin is present, the pressure on the oxidation technology to be selective is not as great as in the later cycles where much less lignin is present to capture the oxidant and protect the cellulosic components. For example one of the early sequences for fully bleached kraft pulp involved CECEHEDED, where E stands for alkaline extraction. The more selective D cycles are used in the later stages of bleaching. The pulp is usually washed between each treatment.

Selectivity for lignin over cellulose oxidation is a vital parameter in pulp bleaching; the higher the selectivity, the longer the cellulose polymer chains in the fibers, the stronger the final paper product. The selectivity can be deduced from pulp viscosity versus kappa number graphs. The kappa number measures the amount of lignin present on the pulp by a TAPPI (The Association of the Pulp and Paper Industry) standardized titration using $KMnO_4$; the % lignin ≈ 0.15 × kappa number. Pulp strength is proportional to the length of the cellulose chains and the viscosity of pulp solutions is used as an indicator of pulp strength. Pulp is dissolved in a cellulose solvent (alkaline cupriethylenediamine, CED) and the viscosity of the homogeneous solution is measured. Thus, oxidative bleaching cycles are optimized in terms of selectivity by finding the conditions that result in the best achievable balance between reducing the kappa to the maximum possible extent while minimizing the loss in the viscosity. Selectivity graphs are important for indicating the potential quality of a bleaching technology; several of them are presented in the section on P_{Fe} technology. One of the remarkable things to note from this brief history of pulp bleaching technologies is that there are no commercialized catalysts. The chemical industry uses catalysis extensively to control selectivity, save energy and improve cost-effectiveness. Bringing the benefits of catalysis to the pulp and paper field is perhaps the principal benefit that green chemists can bestow on this industry.

5.1.3 Environmental factors associated with pulp bleaching

In recent decades, the environmental significance of chlorine-based bleaching has become well recognized.[21] Pulp bleaching by elemental chlorine leads to bleaching by-products, called 'absorbable organic halogen' or 'AOX', that are mostly fragments of degraded lignin that have accumulated chlorine in the delignification process. The name AOX arises because, in one method of study, the chlorinated by-products are collected for analysis by absorption onto activated carbon. Many are toxic. Among AOX compounds, there are 75 mono- or polychlorinated dibenzo-p-dioxins (PCDDs) and 135 mono- or polychlorinated dibenzofurans (Fig. 5.1), each differing from the others in the Cl-substitution pattern. Seventeen of these compounds are considered toxic; each is chlorinated at least in the 2,3,7,8-positions. One in particular, 2,3,7,8-TCDD or 2,3,7,8-tetrachlorodibenzo-p-dioxin, is exceptionally toxic and carcinogenic and is always present when Cl_2 is used in the bleaching of kraft pulp.[10] As a result, government regulations and market pressures in the early 1990s led to the virtual elimination of C bleaching in Sweden, Finland, and, to a large extent,

Canada as well. On November 14, 1997, the United States EPA signed a Cluster Rule mandating that, within three years, the U.S. P&PI should eliminate from effluents dioxin and 12 chlorinated phenolics to non-detectable quantities and should reduce all AOX to specific values. Environmental considerations now dominate strategic thinking in the P&PI. As John W. Creighton Jr., President and CEO of Weyerhausr Co. recently noted,[22] his company is striving to eliminate emissions and discharges to reduce reliance on end-of-pipe treatment, while keeping its capital expenditures in line to maintain adequate profitability. The goals are to achieve full compliance with all regulations while also achieving full cost recovery. How is the industry currently meeting the challenge?

In addition to elemental chlorine, effective delignifying and/or lignin decolorizing oxidants include chlorine dioxide (D), hypochlorite (H), oxygen (O), hydrogen peroxide (P) and ozone (Z). O, P and Z processes are totally chlorine free (TCF). The term 'elemental chlorine free' or 'ECF' bleaching is used for chlorine-based bleaching that does not employ Cl_2, that is H and D bleaching. Among the chlorine-based processes, C bleaching produces by far the worst AOX in terms of the quantity as well as the combined toxicity of the individual components. Among all the alternative ECF and TCF technologies, D bleaching has expanded the most in the last decade[23] such that the term 'ECF' bleaching has become synonymous with D bleaching. D processes are winning not only because of the large improvement for the environment over C bleaching, but also because D bleaching is well understood and superbly effective at achieving high brightness without pulp degradation. Of particular importance, in comparison with all existing TCF technologies, D bleaching is highly selective at low kappa numbers.

Once again we see the importance of selectivity as a practical force underlying decision-making processes, this time essentially dictating the choice of bleaching processes in the P&PI. The P&PI is heavily capital intensive. Only reliable bleaching processes that meet environmental, cost and technical standards are attractive for commercialization. Nevertheless, D bleaching would appear to be less than ideal for the industry. It is one of the most capital intensive of the bleaching technologies. Chlorine dioxide is among the most expensive bleaching chemicals. It is produced in the mill by reduction of chlorate salts. D bleaching is not totally chlorine free. Its use is being expanded with the intention of closing mills to water-borne AOX emissions; such closures become especially difficult and expensive when chlorine in any form is employed. Chlorine-containing effluents cannot be burned in the recovery boiler because they lead to corrosion problems. In addition, there is the potential for chlorinated dioxin production if combustion proceeds in the presence of chlorine, fly ash and certain metal ions.[12] In contrast, TCF effluents from O, P or Z bleaching are generally considered to be compatible with the mill recovery process although there can be a problem associated with insufficient calorific content of bleach plant effluent making it unsuitable for combustion in recovery boilers.

5.1.4 Why TCF bleaching has not been becoming predominant

The most widely adopted TCF technology in terms of pulp volume is O bleaching.[24] The first commercial system started operation in 1970 and total world capacity in 1980 was only about 10 000 tons per day. By 1992, the number of operating systems stood at 155 with a total capacity of about 85 000 tons per day. But TCF technology has not continued to expand significantly after 1994. In contrast, from 1990 to 1995, ECF (D) production grew more than 700% to total 28.5 million tonnes. By 1997, ECF commanded the highest worldwide market share at 50%, totalling 38 million tonnes.[23] For O bleaching, the principal advantages pertain to the environment and to the relatively low chemical costs. However, the principal disadvantages are the high capital costs of O systems and the tendency of O systems to be non-selective, especially at higher degrees of delignification. It has been estimated that the installed cost of an O delignification system with two post-oxygen washing stages lies in the range of $13–26 M depending on the equipment selection and the site. Compared with D bleaching, O bleaching is markedly less effective at removing lignin at low pulp lignin content. Another problem involves a further high capital item, namely, the recovery boiler. Many mills were built with the expectation that the recovery boiler would operate at or near its capacity. Thus, the advantage of sending additional organic material from a TCF bleach plant can often not be gained because additional material from the bleach plant would overload the recovery system. Because O-bleaching is carried out at relatively high temperatures (85–115 °C) and pressures (415–800 kPa), further disadvantages include increased steam costs, higher maintenance costs and an increase in overall process complexity.

Hydrogen peroxide has a dual role in bleaching and delignification of pulp.[25] High brightness can be achieved both by removal of lignin from the pulp or by decolorization of residual lignin in the final pulp through removal of color centers. The primary role of alkaline peroxide today lies in its ability to brighten lignin-rich pulps to high brightness levels of 80–83% without substantial disintegration or dissolution of the lignin. Brightened lignin-rich pulps can be used for short-term products such as newsprint which rapidly yellow on aging. More recently however, Swedish, Spanish and Canadian mills have started using peroxide for a second role, namely extensive delignification, which is possible with prolonged high temperature alkaline peroxide treatment of the pulp. The great advantages of P bleaching are that it has relatively low capital intensity and is easy to use. There have been three big disadvantages. First, as with O bleaching, P bleaching is less effective than D bleaching at low pulp lignin content. Second, hydrogen peroxide has been an expensive oxidant, but this disadvantage has faded as global peroxide production capacity has expanded and the cost of the chemical has dropped; it is currently less expensive by the pound than chlorine dioxide. Third, peroxide delignification is generally quite slow. If the third disadvantage could be overcome by finding a cost-effective, selective, catalytic technology, then peroxide should enjoy greater usage for first and second stage delignification of kraft pulp. If the first disadvantage could also be overcome by catalysis, then peroxide bleaching would become much more important than currently is the case and would perhaps rise to become the dominant bleaching technology.

Ozone bleaching (Z) is highly capital intensive and the process is comparatively complex. Nevertheless, the Z bleaching is fast and the coupling of this advantage to the TCF environmental benefits has induced several companies in the US to work on perfecting Z technology. Impressive facilities have been built in several mills in which ozone is used for initial delignification of kraft pulp followed by several D phases for final delignification and attainment of high brightness.

5.1.5 How TCF bleaching might increase in significance and why it should

ECF (D) bleaching has become the paramount technology because no existing TCF technology presents a sufficiently attractive solution to the integrated needs of the environment and the P&PI. It is important to reiterate that no catalytic activator of a peroxide or oxygen based process has achieved marketing success. In this way, pulp bleaching provides a microcosm of the universal problem in oxidation chemistry. Homogeneous oxidation technology is poorly developed because of a dearth of effective catalysts.[26] Consequently, the abundant natural oxidants, oxygen and hydrogen peroxide, have limited technological significance compared to what is desirable. In the near future, numerous remedies for environmental problems will be developed around catalysts that activate one or other (or both) of these natural oxidants. In the P&PI, where the move to ECF bleaching has significantly improved the industry's environmental performance at great expense, such activators should be developed primarily to increase the flexibility and profitability of papermaking. Further improvements in environmental performance should be regarded as added benefits of the new technologies.

We have argued that broadly useful catalysts are not available because oxidatively robust ligands are rare.[26] We have described a design domain for producing oxidation catalysts that is complementary to that in Nature.[27] In this new domain, we have been developing since 1980 oxidatively robust tetra-amido macrocyclic ligands via an iterative design process in which oxidatively sensitive ligand moieties are identified and replaced. The resulting catalysts are also designed to contain environmentally friendly elements. Recently, this process has led to highly effective catalysts for the bleaching of organic dyes with hydrogen peroxide.[27–30] In the remainder of this chapter, we will summarize the utility of the catalysts in the pulp and paper field as activators of peroxide for bleaching kraft pulp and conclude by introducing results that suggest promise for decolorizing spent mill liquors and for decomposing AOX. The following review is based upon results that have been described elsewhere.[31]

5.2 New activators of hydrogen peroxide

5.2.1 The peroxide activators and the experimental methods for their use

Currently, research is being conducted in several laboratories, including ours, to attempt to obtain viable catalyzed oxygen[32] or hydrogen

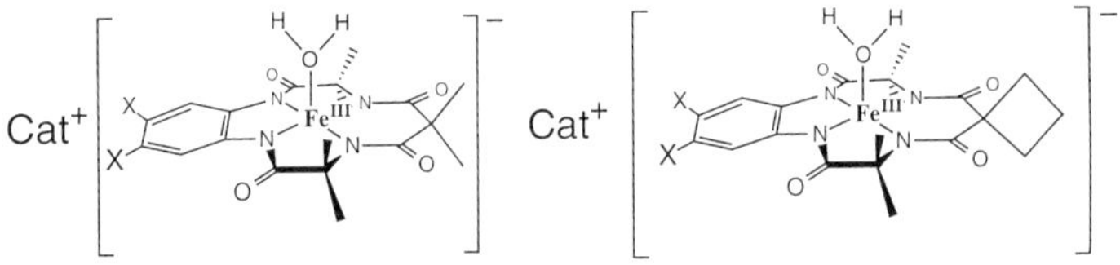

Fig. 5.4 The iron activators of hydrogen peroxide.

1a; X = Cl
1b; X = H
1c; X = OCH$_3$

Cat$^+$ = Li$^+$, [Me$_4$N]$^+$, [Et$_4$N]$^+$, [PPh$_4$]$^+$

2a; X = Cl
2b; X = H

peroxide[33] processes for pulp bleaching. Representative examples of P$_{Fe}$ activators are shown in Fig. 5.4. While all the activators of Fig. 5.4 show significant bleaching behavior, we will focus here on the use of activator **2a**. It is important to note that the chlorine atoms in the structure of **2a** are not necessary for catalytic activity. However, they do facilitate research scale isolation procedures of pure activators.

All experiments were performed using well-washed commercial *Pinus radiata* kraft pulp having a kappa number of 22.5 and a viscosity of 37.4 mPa.s; this origin point is not presented on many of the figures that show selectivity and efficiency measurements. All chemical charges are based on the equivalent weight of oven-dried pulp used in the experiment (the actual pulp used was not dried). Chelation pretreatment was performed for 60 min at 10% consistency using 1.0% EDTA on pulp at pH 5 and at 90 °C.[34] In a typical P$_{Fe}$ run, the pulp (wet pulp corresponding to 5 g of oven dried pulp), NaOH (1–3% on pulp), MgSO$_4$ (0.3% on pulp), and catalyst (quantity as specified), and DTMPA, (diethylenetriaminepentamethylenephosphonic acid, quantity as specified, if added) and water required to give a final consistency of 10% were thoroughly mixed and equilibrated at 50 °C or 90 °C. In all experiments, 0.1 mg of **2a** corresponds to 20 ppm. Hydrogen peroxide was then kneaded into the mixture to begin the reaction. For most treatment conditions, a set of five reactions was performed. After 15 min, one of the reactions was terminated and a further portion of catalyst was kneaded into the remaining four. Addition of further portions of catalyst and removal of one of the remaining reactions continued at 15 min intervals. Thus, for the last treatment, a total of five catalyst additions were made for a total reaction time of 75 min. As indicated in the figure captions, the time between additions was reduced to 7.5 min for one treatment set. To stop the treatment, the reaction liquid and the pulp were separated and the pulp was washed at 1% consistency with distilled water adjusted to pH 5–6. Residual H_2O_2 in the reaction liquid was determined by iodometric titration and the treated pulp was analyzed according to TAPPI standard methods UM246 (micro-Kappa) and T230 (relative viscosity).

Uncatalyzed peroxide treatments were performed at 10% consistency (10% pulp, 90% water and chemicals) using H_2O_2 (1–4%), NaOH (2–3.5% on pulp) and $MgSO_4$ (0.3% on pulp) at 90 °C for the indicated time. Oxygen delignification[35] was performed in a pressure vessel for 60 min at 10% consistency and 100 °C under oxygen (0.7 MPa) in the presence of 0.5–2.5% NaOH and 0.1% $MgSO_4$ on pulp. Chlorine dioxide treatments were performed at 50 °C for 10–15 min and 10% consistency in the presence of NaOH (0.2–0.45% on pulp) using commercial chlorine dioxide solution (2.25–5.63% on pulp as available chlorine). The treated pulp was then extracted at 10% consistency and 70 °C for 60 min with NaOH (2% on pulp).

5.2.2 The control experiments

Control experiments were carried out for P bleaching, O bleaching and D bleaching (Fig. 5.5). All of the data presented in this paper were obtained employing the same sample of pulp. As is well-known for O bleaching,[36] the extent of delignification can be controlled by the quantity of added base. Measurements were taken with 0.5, 1, 1.5 and 2.5% NaOH (on od pulp). For peroxide bleaching, it is known that there is an optimum charge of NaOH as a function of H_2O_2 charge.[25] Control of pH was achieved by adding a charge of 2% or 3% NaOH (on oven dried pulp) at the beginning of the process. The peroxide delignification behavior was also studied with and without addition of metal ion chelating agents: Q = EDTA pretreatment, * = DTMPA cotreatment.

Figure 5.5 illustrates several important points. First, with this softwood pulp, peroxide delignification under appropriate conditions leads to higher selectivity than conventional oxygen delignification. Second, peroxide delignification at 50 °C is ineffective. Third, despite its relatively favorable properties at higher kappa numbers, P delignification is comparatively slow; it takes 6 h to reach the final kappa number of 7.

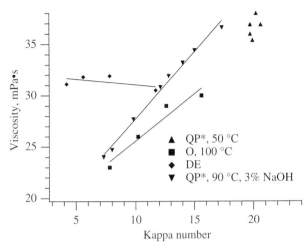

Fig. 5.5 Control reactions: Selectivity (viscosity vs Kappa number) for bleaching of Kraft pulp by H_2O_2 (P, 4%), O_2 (O), ClO_2 followed by alkaline extraction (DE). Q indicates EDTA pretreatment and * indicates treatment in the presence of DTMPA. ■ O delignification – each data point was obtained after 60 min of reaction at 100 °C. NaOH concentrations were 0.5, 1, 1.5 and 2%. ▲ QP* at 50 °C. Data points were taken every 15 min for a total reaction time of 75 min. ♦ DE. ▼ QP* at 90 °C 3% NaOH. The first five data points were taken every 15 min for 75 min and the final three data points were taken after 2, 4 and 6 h.

5.2.3 Catalyzed pulp bleaching at 50 °C

Figure 5.6 presents the selectivity data for the catalyzed peroxide treatments at 50 °C. These 50 °C treatments utilized 1% NaOH; later optimization showed that 1.5% NaOH at 50 ° C leads to slightly better selectivity. The results of the DE and QP* control experiments are also shown on Fig. 5.6 for reference. Noteworthy features of the data are that the catalyzed process produces substantial delignification compared to the uncatalyzed process, where little delignification occurs, that chelation applied before the bleaching process marginally enhances the selectivity, and that a combination of EDTA pretreatment with DTMPA cotreatment significantly improves the selectivity. A second significant parameter of P-based bleaching is the efficiency in terms of the peroxide consumed per unit weight of bleached pulp. Figure 5.7 shows the P_{Fe} efficiency data collected at 50 °C. Because

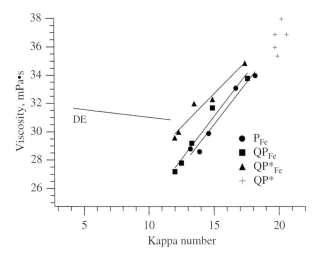

Fig. 5.6 Selectivity (viscosity *vs* Kappa number) for bleaching of Kraft pulp with H_2O_2 (4%) at 50 °C by **2a** (Fe). Q indicates EDTA pretreatment and an * indicates treatment in the presence of DTMPA. In all cases **2a** was added in 0.1 mg quantities and the data taken 15 min after each addition for a total reaction time of 75 min. DE is ClO_2 treatment followed by alkaline extraction.

the uncatalyzed processes do not proceed to a useful degree at 50 °C, the uncatalyzed and catalyzed efficiency data at 90 °C are also presented.

Examination of Fig. 5.7 shows that both the catalyzed and uncatalyzed 90 °C processes are more efficient than QP^*_{Fe} at 50 °C. However, the efficiency is very dependent upon the base concentration and the optimal values are different at 50 and 90 °C. The 90 °C data was obtained with 3% NaOH added at the beginning of the treatment. The optimal base concentration at 50 °C was found to be about 1.5% NaOH where the process has about 85% of the efficiency the QP* (90 °C, 3% NaOH) process. It is remarkable that bleaching can proceed so rapidly and selectively at 50 °C. This discovery suggests that P_{Fe} technology can bring considerable flexibility to bleach plants since a rapid bleaching process at such a low temperature would have minimal capital demands and would save energy compared to existing processes.

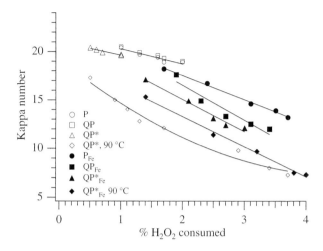

Fig. 5.7 Efficiency of H_2O_2 use (Kappa number vs consumed H_2O_2) for bleaching of Kraft pulp with H_2O_2 (P, 4%) at 50 °C in the absence and in the presence of **2a** (Fe). Q indicates EDTA pretreatment and * indicates treatment in the presence of DTMPA. In the cases of ●, ■, ▲ and ▼, **2a** was added in 0.1 mg quantities and the data taken 15 min after each addition for a total reaction time of 75 min. ○, □, △ and ◇ have no added **2a**. The first five data points were taken every 15 min for 75 min and the final three data points were taken after 2, 4 and 6 h.

5.2.4 Catalyzed pulp bleaching at 90 °C

Figure 5.8 presents the selectivity data for the catalyzed peroxide treatment at 90 °C. Bleaching experiments at this temperature were carried out with chelation treatment because it is well known that such treatment is essential for ensuring the selectivity and efficiency of hot peroxide delignification processes. The results of the DE and QP* control experiments are also shown on the figure for reference.

Several features of the results are noteworthy. First, the addition of **2a** does not significantly alter the selectivity of the delignification process. Second, the presence of **2a** even in very small quantities dramatically enhances the rate of the delignification process. For example the uncatalyzed QP* process takes 6 h to achieve a kappa number of 7.3 and a viscosity of 24.0 mPa.s. The QP^*_{Fe} process with four catalyst additions at 15 min intervals takes 60 min to achieve a kappa number of 7.5 and a viscosity of 25.5 mPa.s. Addition of a fifth aliquot of catalyst at 60 min leads after 75 min to a little change; to a kappa number of 7.3 and a

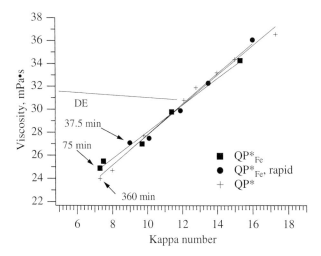

Fig. 5.8 Selectivity (viscosity vs Kappa number) for bleaching of Kraft pulp with H_2O_2 (P, 4%) at 90 °C by **2a** (Fe). Q indicates EDTA pretreatment and * indicates treatment in the presence of DTMPA. ■ **2a** was added in 0.1 mg quantities and the data taken 15 min after each addition for a total reaction time of 75 min. ● **2a** was added in 0.1 mg quantities and the data taken 7.5 min after each addition for a total reaction time of 37.5 min. DE is ClO_2 treatment followed by alkaline extraction.

viscosity of 24.9 mPa.s. The activator additions can be made every 7.5 min. This more rapid addition style returns comparable bleaching selectivity but slightly lower efficiency (Fig. 5.9).

Figure 5.9 presents the efficiency data for the catalyzed peroxide treatments at 90 °C. One observation is particularly noteworthy. The QP* treatment has an efficiency advantage over QP^*_{Fe} in the early stages of the 4% peroxide consumption. However, by the end of the peroxide consumption, the efficiencies of QP* and the QP^*_{Fe} that was allowed to run for 75 min are almost equivalent. The principal advantage of using the activators at 90 °C is the dramatic increase in rate. This can be an important advantage. Mills represent such a great capital investment that achieving the maximum throughput is important for profitability.

It is interesting to consider how these results would appear in a mill process on a per tonne of pulp basis if they proved to be exactly transferable. Consider first the slower catalyst addition style (15 min intervals) at 90 °C. With 40 kg of peroxide per tonne pulp and 80 ppm of

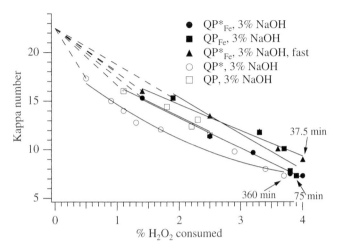

Fig. 5.9 Efficiency of H_2O_2 use (Kappa number vs consumed H_2O_2) for bleaching of Kraft pulp with H_2O_2 (P, 4%) at 90 °C in the absence and in the presence of **2a** (Fe). Q indicates EDTA pretreatment and * indicates treatment in the presence of DTMPA. In the case of ● and ■, **2a** was added in 0.1 mg quantities and the data taken 15 min after each addition for a total reaction time of 75 min. For ▲ **2a** was added in 0.1 mg quantities and the data taken 7.5 min after each addition for a total reaction time of 37.5 min. ○ No added **2a**. The first five data points were taken every 15 min for 75 min and the final three data points were taken after 2, 4 and 6 h. □ No added **2a**. The data points were taken every 15 min for a total reaction time of 75 min.

2a (80 g) added in four portions at 0, 15, 30 and 45 min, after 1 h 68% of the lignin could be removed giving a final kappa number of 7.8 and a final viscosity of 25.3 mPa.s. Consider next the faster catalyst addition style; (7.5 min intervals) at 90 °C. Here 40 kg of peroxide and 100 ppm of **2a** were added in five portions at 0, 7.5, 15, 22.5 and 30 min. After 37.5 min, 60% of the lignin could be removed giving a final kappa number of 9 and a final viscosity of 27.1 mPa.s. If the pulp was then to be fully bleached and taken to full brightness with D technology, P_{Fe} activators would have enabled the rapid removal of approximately half the lignin in a TCF process without compromising the selectivity. Moreover, it might be possible to achieve the lignin removal in, for example, the brown stock holding tank of a mill at ca. 50 °C employing residual heat from the kraft delignification process.

With the current generations of activators, it was initially found that P_{Fe} technology gave superior selectivity and efficiency if the activator was kept at low concentration (ca. 20 ppm or less). This was achieved by sequential additions of the activator to the bleaching media. Thus, most treatments presented here involved four or five activator additions. It was subsequently found that improved mixing of the activator with the pulp prior to addition of peroxide reduced the need for sequential additions. Thus, 0.3 mg (60 ppm) of **2a** was heated for 60 min at 90 °C with the pulp in the bleaching medium minus the peroxide. Peroxide (4% on od pulp) was then added and the bleaching process was allowed to proceed at 90 °C for 45 min giving a kappa number of 11.5 and a viscosity of 31.2 mPa.s. This significant practical feature will be more fully described at a later date.

We know that **2a** and the other activators presented in Fig. 5.4 undergo slow decay in the pulp bleaching process. This is why mixing is so important because catalyst degradation and lignin oxidation appear to be competing processes. By continuing our iterative ligand design procedure, we are working to produce new generations of catalysts that will exhibit even longer lifetimes. With such modified activators, it should be possible to employ P_{Fe} technology to achieve higher selectivity and efficiency with a considerably smaller catalyst charge administered in a single addition.

5.2.5 Preliminary data on the destruction of AOX and the bleaching of effluent color

Two preliminary studies illustrate that P_{Fe} technology has potential applications in the environmental management of mill processes. These areas are AOX destruction and color reduction. A full study of the environmental utility of P_{Fe} technology will be undertaken and these results should be viewed as preliminary leads. However, the promise of the results is supported by the extensive knowledge we have accumulated into the ability of the catalysts to rapidly bleach organic dyes and to oxidize organic matter in general.[27,28,30]

The AOX destruction study was performed using 2,4,6-trichlorophenol (TCP) as a representative compound. In a very thorough study, Meunier *et al.* have reported on the use of water soluble iron phthalocyanine complexes to catalyze the oxidation and mineralization of TCP with hydrogen peroxide.[37,38] In the Meunier process (Fig. 5.10),

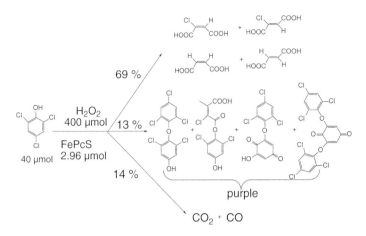

Fig. 5.10 Oxidation products resulting from the reaction of 2, 4, 6-trichlorophenol (TCP) with 400 μmol H_2O_2 in the presence of 2.96 μmol of the water soluble iron complex of 2, 9, 16, 23-tetrasulfophthalocyanine (FePcS). The coupled products are purple, TCP is colorless and the FePcS/H_2O_2 system only partially converts the coupled products to water-soluble derivatives.

significant degradation and mineralization was observed, as well as the formation of coupled phenol products which, it was suggested, are also likely substrates for the oxidation system. We have oxidized TCP with hydrogen peroxide in the presence of activator **1a** and followed the process by UV/Vis spectroscopy. Figure 5.11 shows a stacked plot of the data over time (united digitally with Mathematica to give a surface). This plot reveals that the TCP absorption at 316 nm is bleached rapidly. A band at 516 nm that is indicative of the purple intermediates detected by Meunier initially forms, but then is also completely bleached within 200 s. After 350 s, the spectrum from 350 to 650 nm is essentially bleached.

To test for recalcitrant color destruction, we obtained effluent from a kraft pulp mill. The UV/Vis spectrum of the effluent exhibits a featureless absorbance in the UV which tails into the visible spectrum. The results of a bleaching study are presented in Fig. 5.12. While peroxide by itself actually causes an increase in the color measured at 400 nm which might be due to formation of a finely divided precipitate, the presence of the **1a** causes a gradual bleaching. After 5000 s and two additions of **1a** (2 × 0.14 ppm) with 200 ppm of H_2O_2, 60–70% of the absorbances at 300, 350, 400 and 450 nm had been bleached.

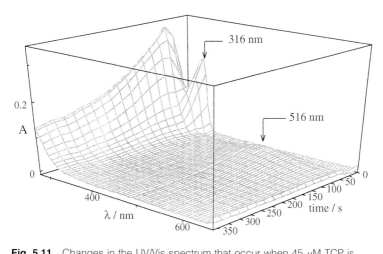

Fig. 5.11 Changes in the UV/Vis spectrum that occur when 45 μM TCP is oxidized in pH 7.4 phosphate buffer by 1 μM **1a** in the presence of 5.4 mM H_2O_2. In this experiment, **1a** and TCP were first combined in the pH 7.4 buffer, then H_2O_2 was added approximately 10 s after the t = 0 scan, and then the spectral changes were monitored every 20 s. TCP absorbance maxima occur at 220, 256 and 316 nm.

5.3 Conclusion

Shifting the elemental balance of technology closer to that which Nature uses for the same or a related process is an important theme for green chemistry. For the sake of the environment, the P&PI has invested substantially in the last decade to replace elemental chlorine as its principal bleaching agent with the ECF alternative, chlorine dioxide. TCF bleaching currently represents a small component of the marketplace. Expansion of TCF bleaching makes the most sense if it is pursued to complement D bleaching in mills in a cost saving manner. Catalysis of P bleaching could bring about such cost savings by leading to a reduction in chlorine-containing effluent, to savings in energy costs and to increased process flexibility, all without slowing mill operations, without demanding significant capital investment and without compromising product quality. The bleaching activators reviewed have the potential to bring these advantages to the industry. Their development has been dependent upon attaining a deep understanding into how

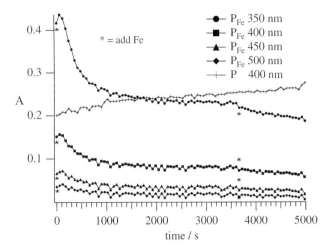

Fig. 5.12 Bleaching of effluent from a kraft mill as monitored by UV/Vis spectroscopy at the wavelengths indicated on the figure. Reaction conditions for ●, ■, ▲, and ◆,: 3 ml of effluent, 0.14 ppm **1a** (added at t = 0 s and t ≈ 3660 s, indicated by *), 200 ppm H_2O_2 (added approximately 10 s after the t = 0 s spectrum), pH adjusted to 10–11 with NaOH, T = 35 °C. Reaction conditions for +: 3 ml of effluent, 200 ppm H_2O_2 (added approximately 10 s after the t = 0 s spectrum), pH adjusted to 10–11 with NaOH, T = 35 °C.

ligands decompose when metal complexes are subjected to oxidizing media.[27–30,39] Our understanding of decomposition processes now allows us to obtain a family of activators with differing finite lifetimes and it has set the stage for expanding the family to include members with longer lifetimes and other selectivity enhancing features. This work should enable chemists and engineers to choose a particular activator to perform a desired function after which it would decompose in the presence of excess oxidant.

P_{Fe} bleaching has the potential to help shift the elemental balance of wood pulp bleaching towards the peroxide (from oxygen) process used by Nature for lignin degradation.[1] Catalytic activation provides the method of choice for achieving reliable processes that do not suffer from unpredictable and incomprehensible instabilities. P_{Fe} technology is clearly not capital intensive and offers considerable flexibility such that each individual mill should be able to tailor it to suit its particular environment. Compared with D bleaching, P_{Fe} bleaching should bring operating cost savings in the energy and effluent sectors. The attractiveness of P_{Fe} delignification will depend on the relative pricing of hydrogen peroxide and chlorine dioxide at any given place and time. P_{Fe} technology promises new possibilities for assisting mills to eliminate recalcitrant effluent problems. Through bleaching processes and numerous other conceivable applications, P_{Fe} technology appears to be fully capable of providing the P&PI with an opportunity to achieve cleaner, more cost-effective papermaking.

Acknowledgement

This research was funded by the National Institutes of Health (GM44867-05) and the National Science Foundation (CHE9612990). JAH acknowledges receipt of an AGMARDT Postgraduate Scholarship and the University of Auckland.

References

1. Tien, M. *CRC Critical Reviews in Microbiology* **1987**, *15*, 141–168.
2. Anastas, P.T.; Warner, J.C. *Green Chemistry: Theory and Practice*; Oxford Science Publications: Oxford, **1998**.
3. Collins, T.J. *Green Chemistry*; in *Macmillan Encyclopedia of Chemistry*; Lagowski, R., ed.; Simon and Schuster Macmillan: New York, **1997** Vol. 2, pp. 691–697.
4. Eisinger, J. *Natural History* **1996**, *105*, 48–53.
5. Needleman, H.L., ed.; *Human Lead Exposure*; CRC Press: Boca Raton, **1992**.
6. Molina, M.J.; Rowland, F.S. *Nature* **1974**, *249*, 810.
7. Farman, J.C.; Gardiner, B.G.; Shanklin, J.D. *Nature* **1985**, *315*, 207–210.
8. AFEAS, *Trends in World Bleached Chemical Pulp Production: 1990–1997*; www.afeas.org/.
9. Ramamoorthy, S.; Ramamoorthy, S. *Chlorinated Organic Compounds in the Environment: Regulatory and Monitoring Assessment*; Lewis Publishers: Boca Raton, **1998**.
10. Berry, R. *Pulp Bleaching and the Environment*. Ch. 3: *Dioxins in Effluents, Pulp, and Solid Waste*; in *Pulp Bleaching: Principles and Practice*; Dence, C.W. and Reeve, D.W., ed.; TAPPI Press: Atlanta, **1996**, pp. 799–820.

11. Budavari, S.; O'Neil, M.J.; Smith, A.; Heckelman, P.E.; Kinneary, J.F., ed.; *The Merck Index*; Merck Research Laboratories: Whitehouse Station, NJ, **1996**.
12. Lenoir, D.; Fiedler, H. *UWSF-Z. Umweltchem. Ökotox.* **1992**, *4*, 157–163.
13. Anastas, P.T.; Williamson, T.C., ed.; *Green Chemistry: Frontiers in Benign Chemical Syntheses and Processes*; Oxford University Press: Oxford, **1998**, pp. 1–26.
14. Anastas, P.T. *Benign by design*; in *Benign by Design: Alternative Synthetic Design for Pollution Prevention*; Anastas, P.T. and Ferris, C.A., ed.; American Chemical Society: Washington, D.C., **1994**, pp. 2–22.
15. Anastas, P.T.; Williamson, T.C., ed.; *Green Chemistry: Designing Chemistry for the Environment*; American Chemical Society: Washington, **1996** Vol. 626.
16. Anastas, P.T.; Williamson, T.C. *Green Chemistry: An Overview*; Anastas, P.T., Williamson T.C. ed.; American Chemical Society: Washington, **1996** Vol. 626, pp. 1–17.
17. Smook, G.A. *Handbook for Pulp and Paper Technologists*; 2nd edn.; Angus Wilde Publications Inc.: Vancouver, **1992**.
18. Dence, C.W.; Reeve, D.W., ed.; *Pulp Bleaching: Principles and Practice*; TAPPI Press: Atlanta, **1996** p. 6.
19. Reeve, D. *Introduction to Pulp Bleaching*; TAPPI Press: Atlanta, **1986**.
20. Reeve, D.W. *Bleaching Chemistry*; in *Alkaline Pulping*; Grace, T.M. and Malcolm, E.W., ed.; Harpell's Press Cooperative: Ste-Anne-de-Bellevue, Quebec, **1989** Vol. 5.
21. Springer, A.M., ed.; *Industrial Environmental Control: Pulp and Paper Industry*; TAPPI Press: Atlanta, **1993**.
22. Pinkerton, J.E.; Abubakr, S.; Meadows, D.G. *TAPPI J.* **1997**, *80*, 243–247.
23. Alliance for Environmental Technology. *Trends in World Bleached Chemical Pulp Production: 1990–1997*; WWW.aet.org/science/trends97.html.
24. McDonough, T.J. *The Technology of Chemical Pulp Bleaching*. Ch. 1: *Oxygen Delignification*; in *Pulp Bleaching: Principles and Practice*; Dence, C.W. and Reeve, D.W., ed.; TAPPI Press: Atlanta, **1996**, pp. 213–239.
25. Lachenal, D. *The Technology of Chemical Pulp Bleaching*. Ch. 6: *Hydrogen Peroxide as a Delignifying Agent.*; in *Pulp Bleaching: Principles and Practice*; Dence, C.W. and Reeve, D.W., ed.; TAPPI Press: Atlanta, **1996**, pp. 347–362.
26. Collins, T.J. *Accounts of Chemical Research* **1994**, *27*, 279–285.

27. Collins, T.J.; Gordon-Wylie, S.W.; Bartos, M.J.; Horwitz, C.P.; Woomer, C.G.; Williams, S.A.; Patterson, R.E.; Vuocolo, L.D.; Paterno, S.A.; Strazisar, S.A.; Peraino, D.K.; Dudash, C.A. *The Design of Green Oxidants; in Green Chemistry: Environmentally Benign Chemical Syntheses and Processes*; Anastas, P.T. and Williamson, T.C., ed.; Oxford University Press: Oxford, **1998**, pp. 46–71.
28. Collins, T.J.; Horwitz, C.P. *Metal Ligand Containing Bleaching Compositions*, U.S. Patent 5,853,428, **1998**.
29. Horwitz, C.P.; Fooksman, D.R.; Vuocolo, L.D.; Gordon-Wylie, S.W.; Cox, N.J.; Collins, T.J. *J. Am. Chem. Soc.* **1998**, *120*, 4867–4868.
30. Collins, T.J.; Gordon-Wylie, S.W. *Long-Lived Homogeneous Oxidation Catalysts*. U.S. Patent No. 5,847,120, **1998**.
31. Collins, T.J.; Fattaleh, N.L.; Vuocolo, L.D.; Horwitz, C.P.; Hall, J.A.; Wright, L.J.; Suckling, I.D.; Allison, R.W.; Fullerton, T.J. *New Efficient Selective TCF Wood Pulp Bleaching, TAPPI Pulping Conference Proceedings, Vol. 3*, Montreal, **1998**, 1291–1300.
32. Weinstock, I.A.; Atalla, R.H.; Reiner, R.S.; Moen, M.A.; Hammel, K.E.; Houtman, C.J.; Hill, C.L.; Harrup, M.K. *J. Mol. Catal. A: Chem.* **1997**, *116*, 59–84.
33. Patt, R.; Mielisch, H.-J.; Kordsachia, O., *Novel Kraft Pulp Bleaching Using Catalyzed Peroxide Treatments, Proceedings of the 1998 International Pulp Bleaching Conference*, Helsinki, **1998**, *1*, 111–117.
34. Basta, J.; Holtinger, E.; Hook, J., *Controlling the Profile of Metals in the Pulp Before Hydrogen Peroxide Treatment, Proc. 6th Int. Symp. Wood Pulping Chem.*, Melbourne, **1991**, *1*, 237–244.
35. Liebergott, N.; van Lierop, B. *Pulp Paper Can.* **1986**, *87*, T300–T304.
36. Liebergott, N.; van Lierop, B.; Teodorescu, G.; Kubes, G.J., *Comparison Between Low and High Consistency Oxygen Delignification of Kraft Pulps, TAPPI Pulping Conference Proceedings*, **1985**, 213.
37. Sorokin, A.; De Suzzoni-Dezard, S.; Poullain, D.; Noël, J.-P.; Meunier, B. *J. Am. Chem. Soc.* **1996**, *118*, 7410–7411.
38. Sorokin, A.; Séris, J.-L.; Meunier, B. *Science* **1995**, *268*, 1163–1165.
39. Bartos, M.J.; Gordon-Wylie, S.W.; Fox, B.G.; Wright, L.J.; Weintraub, S.T.; Kauffmann, K.E.; Münck, E.; Kostka, K.L.; Uffelman, E.S.; Rickard, C.E.F.; Noon, K.R.; Collins, T.J. *Coordination Chemistry Reviews* **1998**, *174*, 361–390.
40. Biermann, C.J. *Handbook of Pulping and Papermaking*; 2 edn.; Academic Press: San Diego, **1996**.

6 Microbial desulfurization of petroleum derivatives

E. D'Addario, C. Colapicchioni, E. Fascetti, R. Gianna, and A. Robertiello

6.1 Historical outlines

Thermal-catalytical processes currently used in the petroleum refining industry need: (1) a large amount of energy necessary to produce the severe conditions of the reaction, (2) increasing quantities of hydrogen used in cracking processes and in reductive reactions for the removal of contaminants such as sulfur and nitrogen contained in the oil and (3) complex and costly procedures connected with catalysts regeneration. For these main reasons the environmental impact of refineries is still rather high notwithstanding many results achieved by the introduction of new procedures of waste minimization and recycling, energy saving and pollution prevention.

Environmentally benign biotechnological processes certainly have the potential to cope with this matter. Unfortunately, up to now biotechnology in the petroleum refining industry is used only for waste water treatment but this picture could change in the near future. In fact, thanks to many research efforts performed in recent years, the desulfurization of petroleum derivatives seems to be on the threshold of application. Historically, research on microbial desulfurization of fossil fuels started in the 1950s and the first patent on the subject was issued in the United States by ZoBell[1] in 1953. In the following two decades, research mainly addressed the bacterial leaching of coal for the removal of pyritic sulfur (for typical work see Capes et al.[2]).

Later on, thermophilic bacteria (i.e. *Sulfolobus acidocaldarius*) for the oxidation of organic sulfur contained in oil refinery waste water, coal and crude oil were considered.[3] In the same period, the potential of different *Pseudomonas* strains able to oxidize organosulfur heterocycles in crude oil and coal, through the carbon-targeted pathway with the formation of water-soluble products (mainly phenols), was examined by Finnerty and Robinson.[4] Particularly, these two authors published a first economic evaluation on biological desulfurization of residual fuel oils at 3.0% sulfur to obtain fuel oils at 0.7% sulfur.[4] The study was based on a number of assumptions such as: the use of a reactor operating at 4–5 h liquid residence time with immobilized whole cells in a packed bed, 1–2 months catalyst half-life, 1 : 1 oil to water ratio, substrate for cell growth at \$65 lb^{-1}. The evaluation was performed in order to determine the values of the main parameters needed to obtain costs of the biological process comparable with those of hydrodesulfurization or the direct purchase of low sulfur petroleum. To meet this target the following requirements were indicated: reduction of the residence time to 1 h maximum, increase of the catalyst half-life by many months, cost of the substrate for cells growth at \$10 lb^{-1} and, no formation of water-soluble organic molecules in order to avoid losses of the heating power associated with sulfur-containing molecules.

On the basis of this last conclusion, most of the research carried out since then addressed the isolation of bacterial strains able to operate the selective removal of sulfur through the cleavage of C–S bonds. In doing this, the oxidative metabolism was generally preferred as the reductive one appeared to suffer: (1) hydrogen consumption, (2) less favorable kinetics and (3) needs of rather costly technologies for the accurate abatement of the produced hydrogen sulfide. In accordance with this view, the oxidative cleavage of C–S bonds for selective sulfur removal will be addressed in the following. More details concerning the reductive cleavage of C–S bonds as well as the oxidative cleavage of C–C bonds are reported by Klein and van Afferden[5] and by Finnerty.[6]

6.2 The biological oxidative cleavage of C–S bonds

The, first pure aerobic, heterotrophic, monoacidophilic soil *Pseudomonas* and *Acinetobacter* species, able to convert thiophenic sulfur to sulfate, were obtained by Isbister and Kobylinski.[7] They observed a promising extent of organic sulfur reduction (47%) in Illinois coal # 6.

However, the most important work concerning pure bacterial cultures with documented desulfurization activity has been carried out by the Institute of Gas Technology (IGT, Chicago, USA). In this institute, two pure cultures identified as *Rhodococcus rhodochrous* and *Bacillus sphaericus* and respectively named as IGTS8 and IGTS9 were isolated in 1988–1991.[8] A metabolic pathway for the conversion of dibenzothiophene (DBT) to 2-hydroxybiphenil (2-HBP) and sulfate through intermediates such as dibenzothiophene sulfoxide (DBT-SO), dibenzothiophene sulfone (DBT-SO_2) and dibenzothiophene sulfonate was also proposed.[8] This pathway, known as 4-S, is certainly the most deeply investigated in the current literature both from a genetic and biochemical point of view. Denome et al.[9] firstly addressed the genetic of IGTS8 showing that the desulfurization genes are located in a 4.0 kb fragment of the plasmid DNA. Later on, Piddington et al.[10] performed a molecular analysis of this fragment revealing that the ability to convert DBT to 2-HBP is expressed by a single operon containing three different genes designated *dszA*, *dszB* and *dszC*. Nucleotide and deduced amino acid sequences were first determined by Piddington.[10] In particular, these determinations revealed that the enzyme encoded by *dszC* converts DBT to DBT-sulfone, and enzymes encoded by *dszA* and *dszB* in concert catalyze the transformation of DBT-sulfone to 2-HBP.

Lei and Tu[11] purified and characterized the enzyme encoded by *dszC* revealing that: (1) this enzyme bonds one flavin mononucleotide or reduced flavin mononucleotide ($FMNH_2$); and (2) the $FMNH_2$ is an essential cosubstrate for the activity of the enzyme itself. These authors, analyzing patterns of products formation under different concentration of $FMNH_2$, indicated that DBT-sulfone is formed through DBT-sulfoxide and that molecular oxygen, rather than water, is used to oxidize DBT. As the enzyme is also able to utilize benzyl-sulfide and benzyl-sulfoxide as substrate, the enzyme itself was identified as sulfide/sulfoxide mono-oxygenase.

In this context, the most comprehensive work on the 4-S pathway was published by Gray et al.[12] in 1996 (Fig. 6.1). These authors discovered a second mono-oxygenase ($DBTO_2$-MO) for the conversion of DBT sulfone to 2-(2-hydroxybiphenyl)-benzenesulfinate, in addition to sulfoxide mono-oxygenase described above. Furthermore, a desulfinase encoded by the gene *dszB* for the final conversion to 2-hydroxybiphenil

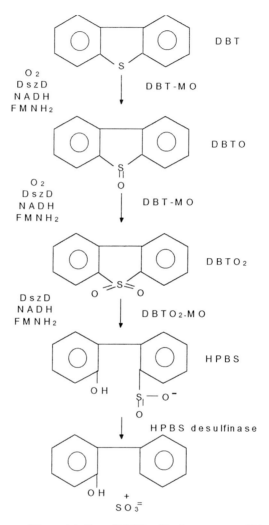

Fig. 6.1 Pathway of the metabolism of DBT by *Rhodococcus sp.* IGTS8 proposed by Gray et al.[12] DBT: dibenzothiophene, DBTO: bibenzothiophene-5'-oxide, $DBTO_2$: Dibenzothiophene-5-5'-dioxide, HPBS: 2-(2-hydroxybiphenyl)-benzenesulfinate, 2-HBP: 2-hydroxybiphenyl, DBT-MO: DBT mono-oxygenase, $DBTO_2$-MO: $DBTO_2$ mono-oxygenase.

was identified. This last step of the pathway was presented as the rate limiting one and it was also found that the two mono-oxygenases require an additional enzyme (DszD). This enzyme, after purification of a single protein and activity measurements in the presence of external cofactors, was found to be a NADH:FMN oxidoreductase. Further investigations into the genetic of IGTS8 have revealed that the region of the DNA 385 bp immediately 5′ to *dszA* contains a promoter and at least three *dsz* regulatory regions.[12] In more detail, it was found that the start of the *dsz* promoter is the G at –46 and that transcription is repressed by sulfate and cysteine but not by dimethyl sulfoxide.[12]

Recently, at the same time as the above fundamental research into IGTS8, a number of different publications related to the isolation of new bacterial strains have appeared. Among these isolates can be mentioned: mesophilic cultures of *Rhodococcus* strain SY1,[13] *Rhodococcus erythropolis* strain D1,[14] *Rhodococcus erythropolis* strains N1-36, N1-43, Q1a-22[15] and *Rhodococcus sp.*[16] formerly identified as *Arthrobacter paraffineus*.[17] Considering that IGTS8 has been recently identified as *Rhodococcus erythropolys*,[17] contrary to previous publications in which this bacterium was indicated as *Rhodococcus rhodochrous* (see for instance Kilbane and Bielaga[8]) all the known mesophilic isolates can be considered to be very similar, and probably they have the same pathway of sulfur removal.

The same metabolism of sulfur removal could probably also take place in thermophilic conditions. In fact, very recently, two new strains identified as *Paenibacillus* and able to remove sulfur from DBT and DBT methyl derivatives in a C–S-bond-targeted fashion with the formation 2-HBP and sulfates have been reported.[18]

In 1994, in Eniricerche laboratories a new bacterial strain able to accomplish the desulfurization according to the 4-S pathway was isolated. After preliminary taxonomic measurements the strain, named DS7, was tentatively identified as *Arthrobacter sp.* and deposited at CBS (Centralbureau Voor Schimmelcultures, Baarn, The Netherlands). Direct chromosomal DNA sequencing revealed that DS7 is identifiable as *Arthrobacter sp.* at 90–95% and as *Rhodococcus sp.* at 100%. Further measurements are still in progress to accomplish the definitive identification of the strain. Resting DS7 cells showed encouraging performances in the desulfurization both of DBT dodecane model system and of some petroleum fractions.[19,20]

6.3 Biodesulfurization in the oil refining industry

According to Monticello[21] biodesulfurization in the oil refining industries presents opportunities for the treatment of the following streams: (1) residual fuel oil, (2) gasoline from Fluid Catalytic Cracking (FCC) and (3) middle distillates. The first option is very stimulating because of the higher price and more favorable market of low sulfur fuel oil, but it requires levels of efficiency hardly compatible with biodesulfurization. The potential application of biological processes for the desulfurization of gasoline from FCC units is very stimulating, mainly considering that conventional hydrodesulfurization of FCC distillates allows olefins saturation and, as a consequence, octane number is reduced. However, this application also appears rather difficult as low molecular weight and aromatic hydrocarbons tend to denaturate bacterial cells. Therefore, nowadays, most of the research efforts addresses the desulfurization of middle distillates. This is due not only to reasons related to the performance of the biocatalyst but also on the basis of the general aspects listed below.

1. Consumptions of mid-distillates (kerosene, diesel and heating oil) are constantly increasing. Halbert *et al.*[22] indicate, for instance, that the percentage of middle distillate consumption versus total crude oil, from 1971 to 1993, rose from 18 to 35% and, by the year 2010, it is expected to increase up to 40%.
2. The technologies (mainly Fluid Catalytic Cracking, FCC) most widely used for increasing distillate production give streams at higher sulfur content (together with nitrogen and aromatics) compared to straight run distillates.[22]
3. The connection between sulfur content and particulate matter emissions in diesel oil is generally accepted. Particularly, the contribution of sulfate (obtained after partial oxidation of SO_2 produced in diesel engines) to particulate emissions has been predicted at 0.067 and 0.004 g kwh^{-1} respectively for fuels at 0.3 and 0.02% sulfur.[23]
4. Under the pressure of environmental concerns, very strict diesel sulfur limits have been promulgated (0.05% in USA and EU, 0.001% Sweden, class 1[22]) and more strict specifications have been proposed

in the European Community (350 ppm by the year 2000 and 50 ppm by the year 2005[24]).
5. Di-beta-substituted dibenzothiophenes are recalcitrant in hydrosulfurization processes[22] while they present a promising biological reactivity because of the absence of phenomena of steric hindrance.[17]

On this basis, in Eniricerche the DS7 bacterium was extensively tested for the deep desulfurization of diesel oil and some technical–economic projections have been carried out in order to determine bottlenecks of the connected biodesulfurization process. In the following, the main results concerning our work on diesel oil as well results of some initial tests on gasoline from FCC units are presented and discussed.

6.4 Experiments of biodesulfurization

6.4.1 Operational

6.4.1.1 Bacterial growth and testing of precultivated cells

DS7 cells were cultured discontinuously in 2-l flasks using the following medium (values in g l^{-1}): KH$_2$PO$_4$ 10.0, NH$_4$Cl 2.0, MgCl$_2$ · 6H$_2$O 0.2, CaCl$_2$ 0.02, FeCl$_3$ 0.01, DBT 0.02, 95%, C$_2$H$_5$OH 8 ml l^{-1}. Cells were harvested by centrifugation, washed twice with 0.1 M phosphate buffer at pH 7.0 and used for desulfurization experiments.

Cells harvested at different times of the bacterial growth were used for specific activity measurements. Experiments were performed at 30 °C using 1 mM DBT in 0.1 M phosphate buffer at pH 7.2. In order to favor DBT dissolution, ethanol at 0.25% v/v was added in the reaction mixture. Cells were used at 1 mg dw ml^{-1}. Experiments into activity measurements lasted 2 h. Pregrown cells were also tested in model systems. For this purpose, an oily phase containing 2 mM DBT in dodecane was prepared. This phase was used as described in the following for diesel oils.

6.4.1.2 Diesel oils

Three samples of diesel oils were tested. The first two were kindly furnished by Agip-Petroli, from Livorno (sample 1) and Gela (sample 2) refineries. Sample 3 was purchased nearby at a local gas station. Initial total sulfur concentration (S$_0$) in sample 1 was 1500 ppm w/w and in samples 2 and 3 it was 400 ppm w/w. Before use, sample 1 was diluted with n-dodecane to reduce S_0 to 400 ppm w/w.

Experiments were performed at 30 °C in the presence of equal volume of phosphate buffer. Cells were used at 20 mg dw ml^{-1} of reaction mixture. All the experiments were performed in 50 ml rubber stoppered vials containing oxygen gas at 1.5 bar. Vials were placed into a water bath and shaken at 300 strokes per min.

In order to improve sulfur removal, diesel oil was also tested by changing cells contained in the reaction mixture. In doing this, after 3 h reaction time, exhausted cells were removed by centrifugation at 6000 × g and fresh cells were added to the vial content.

6.4.1.3 Gasoline from FCC

Precultivated cells were also tested at laboratory scale on distillates obtained by fractionating a sample of gasoline from a typical FCC unit.

The gasoline was kindly furnished by Agip-Petroli and it contained 1200 ppm total sulfur. All the experimental conditions used to test gasoline distillates were equal to those previously indicated for diesel oils.

6.4.1.4 Analysis

DBT and intermediates of reactions for specific activity measurements were extracted with ethylacetate and analyzed via HPLC (Reversed Phase 218-TP 5415 Vydac column).

Organosulfur molecules in diesel oil samples were analyzed on 5890 HP-GC equipped with a 5921-A Atomic Emission Detector (AED). A 30 m long (0.25 mm 5D) DB1-JWS capillary column was used. The following heating program was used: 2 min at 60 °C, 5 °C min^{-1} up to 280 °C, 10 min at 280 °C. Injector and transfer line temperature were 280 °C.

Samples were diluted with dichloroethane and injected in the presence of dibutildisulfide as internal standard. Helium at 60 °C was used as carrier (1 ml min^{-1}, P = 1 bar).

The AED detector was operated as follows: cavity temperature 180 °C, water temperature 66 °C, spectrophotometer purge flow: nitrogen at 2 ml min^{-1}, make-up: helium at 75 ml min^{-1}, ultimate analysis: carbon at 343 nm, sulfur at 361 mm, scavenger gases: hydrogen 90 psi, oxygen 50 psi.

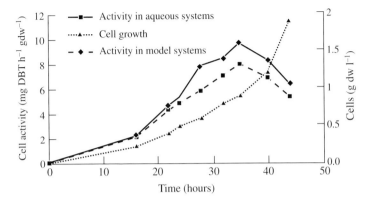

Fig. 6.2 Bacterial growth and cell activity. (Activity measurement: T = 30 °C, pH = 7.2, 1 mg dw l^{-1}, DBT 1 mM, ethanol 0.15% v/v).

6.4.2 Results

6.4.2.1 Bacterial growth and cells activity

Maximum specific activity is given by cells harvested at the beginning of the exponential growth (see Fig. 6.2). At this stage of bacterial proliferation cells concentration was approximately 1 g dw l^{-1} and specific activity of cells in the aqueous and model systems approached 8 and 10 mg DBT h^{-1} g dw^{-1} respectively. The higher activity observed with model systems can be related to: (1) the good dispersion of the oily phase into the aqueous phase which occurs in the absence of external emulsifiers, contrary to different isolates that need this additive as reported by Monticello et al.[21] and (2) the higher solubility of DBT and its intermediates and products of reaction in oily phases and to the consequent easier transport phenomena into dodecane. This last aspect applies particularly to 2-HBP which tends to inhibit cell activity[20] and, in the case of aqueous system, tends to adsorb on the wall of bacterial cells.

A behavior similar to the one shown in Fig. 6.2 was observed for all the numerous experiments performed in our laboratories. However, probably because of phenomena related to plasmid stability, cell activity was not constant. This parameter ranged from 3.5 to 12 mg DBT h^{-1} g dw^{-1}. Experiments aimed at encompassing such a drawback are currently in progress. For this purpose, we are following r-DNA approaches (transferring of genes encoding for enzymes with desulfurizing activity from plasmid into chromosomal DNA) as well as fermentative techniques (selection of optimal operating conditions of continuous cultures).

6.4.2.2 Desulfurization of diesel oil

An initial test with diesel oil was carried out using cells collected from a typical culture. Their specific activity was 5.2 and 6.5 mg DBT h^{-1} g dw^{-1} respectively in aqueous phase and model systems. The obtained desulfurization profiles are showed in Fig. 6.3. As shown in this figure, maximum observed sulfur removals were: 67 (sample 1), 40 (sample 2) and 29% (sample 3). The poor desulfurization of this last sample can be related to the presence of additives (lubricants, antifoaming and anticorrosive agents) contained in commercial diesel oils which negatively affect the performances of the biological reaction. This was noticed mainly in terms of a poor distribution of the oily phase into the aqueous phase, contrary to experiments with sample 1 and 2 which, under typical reaction conditions (shaking at 300 strokes per min), gave quite stable microdispersions.

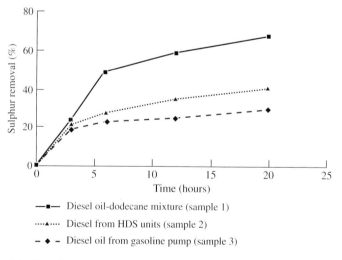

Fig. 6.3 Desulfurization of diesel oil at 400 ppm S. (T = 30 °C, cells 20 mg dw 1^{-1}, oil/water 1:1 v/v).

Table 6.1 Sulfur containing molecules in diesel oils from hydrodesulfurization (HDS) units.

Molecules	Sample 1 (%)	Sample 2 (%)
C_2-alkyl-benzothiophenes	3.2	–
C_3-alkyl-benzothiophenes	1.8	–
C_4-alkyl-benzothiophenes	2.5	–
dibenzothiophene	2.9	–
methyldibenzothiophene	14.7	0.5
C_2-alkyl-dibenzothiophenes	47.6	9.2
C_3-alkyl-dibenzothiophenes	27.3	53.1
C_4-alkyl-dibenzothiophenes		37.2
Total	100.0	100.0

Sample 1: Diesel oil at 400 ppm S obtained after diluting with n-dodecane an original sample 1.500 ppm S from conventional hydrodesulfurization. **Sample 2**: Diesel oil from hydrodesulfurization units able to give products at total sulfur lower than 500 ppm.

Differences of performances between the first two samples are probably due to the presence of different categories of sulfur molecules. Sample 1 originated from a conventional hydrodesulfurization unit able to produce diesel oil at around 1500 ppm total sulfur. Such a sample contained mainly C_2 and C_3-dibenzothiophene alkyl derivatives (approx. 85%, see Table 6.1) and lower amount of dibenzothiophene and methyldibenzothiophenes (approx. 15%). Sample 2 originated from HDS units able to produce diesel oil with residual total sulfur at concentration less than 500 ppm as required by the current Italian standards for city fuels. In these units, process severity was increased and residual organosulfur molecules consisted of negligible amounts of methydibenzothiophene, a certain portion of C_2-alkyl-dibenzothiophenes and, mainly, C_3-C_4-alkyl-dibenzothiophenes. According to our experience, reactivity of precultivated cells towards different categories of organosulfur molecules is in the order: dibenzothiophenes > C_1 dibenzothiophenes > C_1 benzothiophenes > C_2-C_4 benzothiophenes > C_2-C_4 alkyl-dibenzothiophenes. For these reasons, sulfur in sample 2 was removed to a lower extent.

Data presented in Fig. 6.3 indicate that each of the three tested samples were desulfurized to an extent far below the limit proposed by the European Community by the year 2000 (350 ppm) but the standard indicated by the year 2005 (50 ppm, ultra-deep desulfurization) was

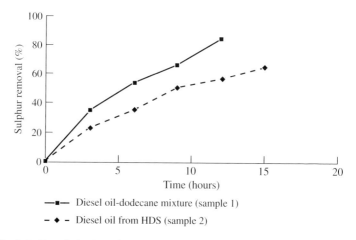

Fig. 6.4 Desulfurization of diesel oil at 400 ppm S with cells renewed each 3 h (T = 30 °C, cells 20 mg/dw 1^{-1}, oil/water 1:1 v/v).

reached in none of the cases. To achieve this target, cells were renewed after 3 h reaction time. According to this approach, cells were changed four and five times respectively in sample 1 and 2. Results obtained from these experiments are showed in Fig. 6.4. As shown in this figure, sample 1 was desulfurized to an extent very close to the target for the ultra-deep desulfurized diesel oil. After four reaction batches, residual total sulfur concentration in sample 1 was about 50 ppm. Fig. 6.4 confirms that desulfurization of sample 2 is rather difficult. In fact, after five reaction batches, residual sulfur concentration was rather higher than in sample 1 (180 versus 50 ppm).

The encouraging desulfurization obtained for sample 1 is showed in Fig. 6.5 in which GC-AED profiles before and after desulfurization are presented. This figure shows that residual organosulfur molecules are basically C_3 and C_4 dibenzothiophenes (retention time > 32 min) confirming the lower reactivity of this molecular class.

6.4.2.3 Gasoline from FCC units

Four fractions of distillates were collected after atmosphere distillation of gasoline. Total sulfur in each of these fraction as well as distillation yields are given in Table 6.2.

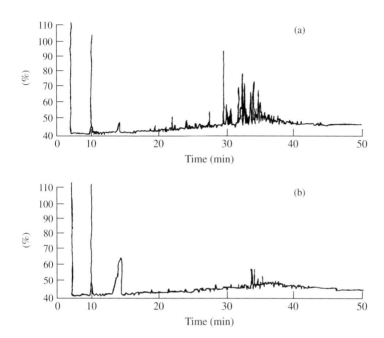

Fig. 6.5 GC-AED profiles of diesel oil at 400 ppm S (sample 1) before (A) and after biodesulfurization (B).

Desulfurization tests were performed using only fraction 2, 3 and 4. The first fraction was considered unusable because of the high content of volatile hydrocarbons which tend to damage bacterial membranes and to denature enzymes catalyzing the reaction of sulfur removal.

Table 6.2 Fractions obtained from atmosphere distillation of FCC gasoline at 1200 ppm total sulfur.

Fraction	Distillation range (°C)	Yield (w/w)	Sulfur concentration (ppm)	Sulfur distribution (%)
1	< 120	57.8	442	21.5
2	120–150	14.1	1600	19.0
3	150–180	13.4	1540	17.4
4	> 180	14.7	3420	42.5
Total	–	100.0	–	100.0

Table 6.3 Initial content of organosulfur molecules in distillates from FCC gasoline (see Table 6.2) and total sulfur removal.

Organosulfur molecules	Fraction 3 (%)	Fraction 4 (%)
Tetrahydrothiophene	0.3	–
C_2-alkyl-thiophenes	18.7	–
C_3-alkyl-thiophenes	61.4	2.3
C_4-alkyl-thiophenes	19.1	23.2
Benzothiophene	0.5	43.7
Alkyl-benzothiophene	–	30.8
Total	100.0	100.0
Total sulfur removal (%)	0.8	7.2
Total sulfur removed (mg)	12.3	246

This assumption turned out to be correct as desulfurization experiments showed that in fraction 2 sulfur removal was undetectable. Table 6.3 shows that desulfurization was almost negligible for fraction 3 and rather low for fraction 4.

Low sulfur removal observed for all of the tested fractions has to be connected again to the poor reactivity of light organosulfur molecules contained in these distillates (alkyl-thiophenes and benzothiophenes, see Table 6.3, in contrast to diesel oils which contain mainly dibenzothiophenes and alkyl derivatives, see Table 6.2).

According to these finding, biological desulfurization of gasoline appears more difficult than middle distillates and it certainly needs longer-term research. Such research should start addressing heavy fractions (> 180 °C) from gasoline distillation and should focus on genetic engineering aspects aimed at improving cells solvent tolerance. Later on, more complex aspects related to the treatment of the whole stream from FCC units could be examined.

6.5 Summary and conclusions

Much progress have been achieved recently on the microbial desulfurization of fossil fuels. This is particularly true for bacteria able to effect the selective removal of sulfur from organosulfur heterocycles through the oxidative cleavage of C–S bonds (4-S pathway). Progress has mainly concerned fundamental research. In fact, many aspects related to the mechanism of the biological reaction, as well as to the genetics of bacteria capable of operating according to this mechanism, have been reported.

Up to now, the most favorable application for this kind of biodesulfurization appears to be the treatment of diesel oils at a maximum 500 ppm S currently produced in the oil refining industry, in order to reduce sulfur down to the legal standard proposed by the European Community by the year 2005 (max. 50 ppm S). Remaining potential applications, such as the desulfurization of high sulfur fuel oils and gasoline from FCC units, need longer-term research. For these areas, research aimed at: (1) increasing solvent tolerability of cells, (2) improving kinetics and life of the biocatalyst, (3) favoring the dispersion of cells into the bulk of heavy oil phases and (4) development of proper bioreactors should be addressed. Clearly, these are very challenging tasks which require extensive research efforts based on genetic engineering, biochemistry and biotechnological subjects.

Less intensive progress is needed for the ultra-deep desulfurization of diesel oil. In this case, it has been estimated that in order to get costs of biodesulfurization comparable with those of new hydrodesulfurization processes able to remove sulfur down to 10–50 ppm, precultivated cells have to express their initial maximum activity for at least 100 h. Therefore, our current research is mainly addressing subjects concerning stability and life of the catalyst. A rigorous comparison between our isolate and those mentioned in the current literature is not possible because of different testing conditions and procedures used by various authors. However, in the case of diesel oil, Johnson[25] reported kinetics similar to those presented in this work for DS7. Nevertheless, beyond the published data, it has to be considered that probably biocatalysts with satisfactory industrial performances have already been developed. In fact, Rhodes[26] reported many initiatives at pilot scale for the desulfurization of diesel fuel and, very recently, Cole[27] announced that a first commercial unit with a capacity of 5000 b d^{-1} of middle distillates will be constructed nearby the Valdez (AK, USA) refinery.

References

1. ZoBell, C.E. (1953). U.S. Patent 2,641,564, June 9.
2. Capes, C.E., Mc Ilhinney, A.E., Sirianni A.F., and Puddington J.E. (1973). *The Canadian Mining and Metallurgical Bulletin*. Nov. 1973, p. 89.

3. Kargi, F. and Robinson, J.M. (1984). *Biotechnology and Bioengineering*, **26**, 687.
4. Finnerty, W.R. and Robinson, M. (1986). *Biotechnology and Bioengineering Symp.* N. 16, p. 205.
5. Klein, J., van Afferden, M., Pfeifer, F., and Schacht, S. (1994). *Fuel Processing Technology*, **40**, 297.
6. Finnerty, W.R. (1993). Proceedings of the Symposium on Bioremediation and Bioprocessing. 205th National Meeting, American Chemical Society, Denver, CO, March 28–April 2, 1993, p. 283.
7. Isbister, J.D. and Koblylinski, E.A. (1985). *Coal Science and Technology*, **9**, 627.
8. Kilbane, J.J. and Bielaga, B.A. (1991). DOE/PC/88891-T1 (DE92012262), Final Report, Dec. 1991.
9. Denone, J.A., Olson, E.S., and Young, K.D. (1993). *Applied and Environmental Microbiology*, **59**, 9, 2837.
10. Piddington, C.S., Kovacevich, B.R., and Rambosek, J. (1995). *Applied and Environmental Microbiology*, **61**, 2, 468.
11. Lei, B. and Tu, S.C. (1996). *Journal of Bacteriology*, **178**, 19, 5699.
12. Gray, K.A., *et al.* (1996). *Nature Biotechnology*. **14**, 1750.
13. Omori, T., Saiki, Y., Kasuga, K., and Kodama, T. (1995). *Biosci. Biotech. Biochem.* **59**, 1195.
14. Izumi, Y., Ohshiro, T., Ogino, Y.H.H., and Shimano, M. (1994). *Applied and Environmental Microbiology*, **60**, 1, 123.
15. Wang, P. and Krawiec, S. (1994). *Archives Microbiology*, **161**, 266.
16. Denis-Larose, C. *et al.* (1997). *Applied and Environmental Microbiology*. **63**, 7, 2915.
17. Lee, M.K., Senius, J.D., and Grossman, M.J. (1995). *Applied and Environmental Microbiology*. **61**, 12, 4362.
18. Konishi, J. *et al.* (1997). *Applied and Environmental Microbiology*, **63**, 8, 3164.
19. D'Addario, E. (1996). *Proceedings of the Symposium AAA Biotech*. Ferrara (Italy). October 1996, Vol. IV, p. 139.
20. D'Addario, E., *et al.* (1997). *Proceedings of the 15th World Petroleum Congress*, Beijing (China), 12–16 Oct. 1997. Topic 17. J. Wiley and sons.
21. Monticello *et al.* (1994). U.S. Patent 5, 358, 870, October 25.
22. Halbert, T., Anderson, G., and Markley, G. (1997). *Proceedings of the 15th World Petroleum Congress*, Beijing (China), 12–16 Oct. 1997. Topic 9. J. Wiley and sons.

23. Booth, M., Reglitzky, A., Pinclin, R. (1997). *Proceedings of the 15th World Petroleum Congress*, Beijing (China), 12–16 Oct. 1997. Review and Forecast Paper N. 11. J. Wiley and sons.
24. Louvel, J. (1997). *Diesel Market Developments in Europe*. WEFA Conference. Lisbon, 20 June 1997.
25. Johnson, S.W. (1995). *American Chemical Society Symposium*, Dallas Oct. 1995. SPE 30671.
26. Rhodes, K. (1995). *Oil & Gas Journal*, May 15, p. 39.
27. Cole, C. (1998). *Octane Week*, March 16, p. 3.

7 Cyclohexane oxygenation with inorganic photocatalysts

Andrea Maldotti, Rossano Amadelli, Alessandra Molinari, and Vittorio Carassiti

Introduction

The search for new catalysts capable of performing oxidations of alkanes in mild temperature and pressure conditions has been attracting the interest of chemists in view of possible applications in fine, as well as in industrial, chemistry. In this framework, the use of environmentally friendly molecular oxygen as oxidizing species, and sunlight, which represents a totally renewable source of energy, is particularly relevant to realizing innovative and economically advantageous processes which conform to pollution prevention (green chemistry).

In the present article we report on the photoinduced oxygenation of cyclohexane under aerobic conditions, at room temperature and atmospheric pressure. Cyclohexane is a very simple alkane, particularly suitable for obtaining information of general interest on the mechanism of hydrocarbon oxofunctionalization; but the main reason why we focus our attention on this substrate is that its oxidation products, cyclohexanol and cyclohexanone, are interesting precursors in the synthesis of chemicals and in the production of nylon, adipic acid, nitro-cellulose lacquers, celluloid, artificial leather and printing inks.

We have investigated the following inorganic photocatalysts: iron (III) prophyrin complexes, photosensitive oxides and polyoxotungstates. The first class of compounds can be considered biomimetic as they are model systems of natural oxygenases.[1,2] Photosensitive inorganic oxides such

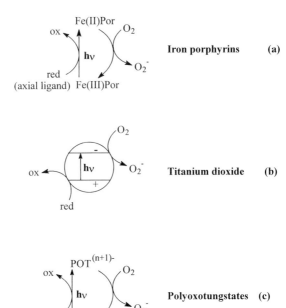

Fig. 7.1 Common redox processes of the inorganic photocatalysts employed.

as titanium dioxide represent interesting photocatalysts mainly because they are stable, inexpensive and can be easily prepared in a high area form. Our interest in polyoxotungstates is twofold. First, they are able to undergo photoinduced multielectron transfer, second they can be considered as soluble models of metal oxide surfaces.

There are close relationships in the photochemical behavior of the above mentioned compounds under aerobic conditions. In the case of iron porphyrins (Fig. 7.1a), irradiation in axial-ligand-to-metal charge transfer (LMCT) bands induces the reduction of Fe(III) to Fe(II) and the oxidation of the ligand to radical species.[3,4] The starting oxidized Fe(III) porphyrin is regenerated as a consequence of the very fast reaction of the ferrous complex with O_2, which consequently undergoes a monoelectronic reduction.[5,6]

Figure 7.1b shows that photoexcitation of TiO_2 at wavelengths corresponding to the band gap energy values leads to charge separation.

Electron are promoted to the conduction band (e⁻(Ti^{III})) and holes are left in the valence band. Particularly in the case of dispersed systems, the separated charge will recombine unless either electrons or holes are efficiently scavenged by oxidants and reductants.[7,8] A large variety of oxidation processes has been investigated on dispersed TiO_2, involving inorganic and especially organic species.[9–11] It is generally accepted that the role of O_2 is not just that of scavenging the photogenerated electrons, but its reduction products O_2^- and H_2O_2 take part effectively in the oxidation process of the substrate.[12,13]

The proposed photochemical pathway of the polyoxotungstate action (Fig. 7.1c) involves the absorption of light by the ground state, producing a charge-transfer-excited state. This process leads to the reduction of the polyoxotungstate with the formation of well characterized blue compounds.[14] A photocatalytic activity of polyoxotungstates has been demonstrated in the oxidation of a variety of organic compounds. The key step in the photocatalytic cycle under aerobic conditions is the reoxidation of the catalyst through a mechanism whereby the one electron oxidation of the photoreduced polyoxotungstate by O_2 is likely to be involved.[15,16] A common characteristic of the described photocatalysts is their ability to generate reactive intermediates in a cyclic way, provided that the initial oxidation state of the metal center can be restored after the primary photoredox reactions. In particular, the possibility of simultaneously oxidizing stable molecules to radical species and of inducing the reductive activation of molecular oxygen can open the way to further interesting pathways in the photocatalytic oxygenation of organic compounds under aerobic conditions. In the following, some of the most significant results regarding the use of these classes of compounds in the oxygenation of cyclohexane are reported.

7.1 Fe(III) porphyrins

7.1.1 Photo-oxidation processes

Iron porphyrins have been found to be biomimetic catalysts for alkane hydroxylation by O_2 with consumption of a stoichiometric amount of a reducing agent according to the mono-oxygenation Equation (1).[17] Moreover, iron porphyrins are capable of catalysing the oxidation of alkanes to alcohols and ketones at 80 °C under an oxygen pressure of

10 atm.[18] In this framework, it appears of particular interest to find catalytic systems able to perform alkane oxidation by O_2 itself under mild conditions and in the absence of a chemical reducing agent.

$$RH + O_2 + 2e^- + 2H^+ \rightarrow ROH + H_2O \quad (1)$$

In recent years, a number of authors described catalytic systems based on the use of photoexcited Fe(III) porphyrins that can induce hydrocarbon oxygenation under aerobic conditions.[19,20] Along this research direction, we have investigated the oxidation of cyclohexane using the *meso*-tetraaryl iron porphyrin reported in Fig. 7.2.[21,22] It is known that the halogen substituents in the *meso*-aryl groups provide a steric protection of the porphyrin ring against radical induced oxidative degradation of the complex during catalytic processes, and prevent the formation of μ-oxo dimers.[23] Photoexcitation of the iron porphyrin complex in catalytic amount at 350–400 nm induces the oxidation of cyclohexane to cyclohexanone and cyclohexanol by molecular O_2, under mild conditions (22 °C; 760 Torr of O_2) and without the consumption of a reducing agent (Table 7.1). The quantum yield values

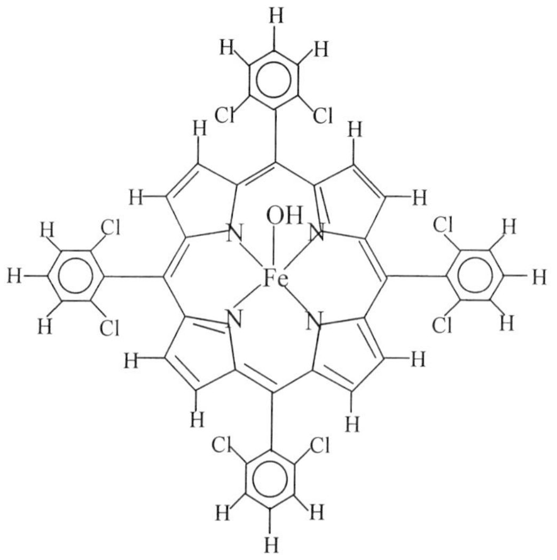

Fig. 7.2 Structure of the substituted iron porphyrin investigated.

Table 7.1 Quantum yields[a] for the Fe(III)TDCPP(OH) catalyzed photo-oxidation[b] of cyclohexane depending on the environment polarity.

Reaction environment	Quantum yields $\times 10^3$	
	$C_6H_{10}O$	$C_6H_{11}OH$
In neat C_6H_{12}	1.6 ± 0.2	< 0.1
In C_6H_{12} / CH_2Cl_2 40/60 (v/v)	1.6 ± 0.2	2.5 ± 0.3
In C_6H_{12} / CH_2Cl_2 40/60 (v/v) + pbn	< 0.1	1.4 ± 0.2

a) Ratio between the moles of oxidized cyclohexane and the moles of the absorbed photons.
b) irradiations were carried out at 22 ± 1 °C, at λ > 350 nm for a period of 60 min. Initial concentration of Fe(III)TDCPP(OH): 2×10^{-5} mol × dm^{-3}.

range from 1×10^{-3} to 5×10^{-3} and about 100 equivalents of oxidized cyclohexane per equivalent of porphyrin consumed are obtained.

The primary photoprocess consists of the homolytic cleavage of the Fe^{III}-OH bond with the formation of OH• radicals (Fig. 7.3 step a). We have evidence for the formation of these intermediates by an ESR spin trapping investigation.[22] This technique is based on the ability of some molecules, such nitrones, to trap radicals to give paramagnetic nitroxides which are stable enough to be studied by ESR techniques. For example,

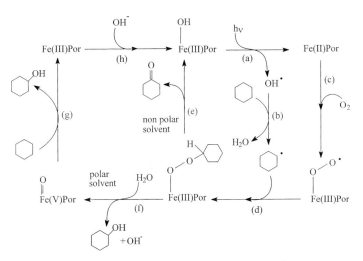

Fig. 7.3 Possible mechanism of photocatalytic oxygenation of cyclohexane by iron(III) porphyrin complexes.

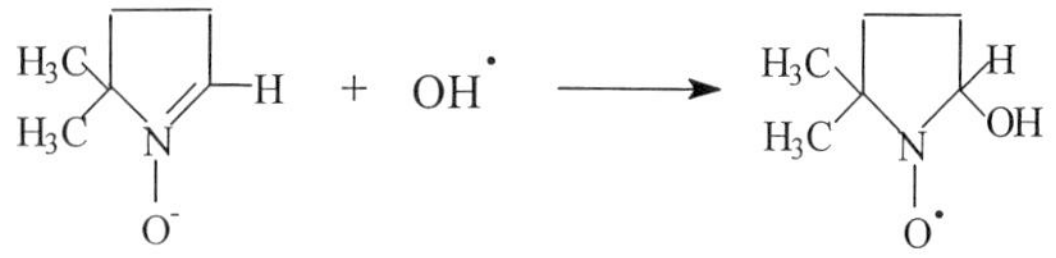

Fig. 7.4 Trapping of photogenerated OH• radicals by DMPO.

5,5-dimethylpyrroline N-oxide (DMPO) is able to trap (Fig. 7.4) the photogenerated OH• radicals to give a nitroxide exhibiting a typical 1 : 2 : 2 : 1 pattern of four lines which can be interpreted as equivalent hyperfine splitting from both β-hydrogen and the nitroxide nitrogen ($a_N = a_H = 14.8$ G). This radical scavenger was chosen on the basis of the conclusions of an investigation[23] where we observed that different nitroxides behave differently in systems of photoexcited iron porphyrins containing bulky ring substituents. In particular, more bulky nitroxides such as phenyl-*tert*-butyl nitrone (PBN) are not able to trap the OH• radical formed in the primary photoreaction, but they can only trap free radicals diffusing away from the first co-ordination sphere of iron.

7.1.2 Effect of the reaction medium

The polarity of the solvent has a remarkable effect on the chemioselectivity of the photo-oxidation process that follows the primary photoprocess described previously. In pure cyclohexane, we observe the selective formation of cyclohexanone from cyclohexane, (Table 7.1) with the stoichiometry of Equation 2. The mechanism that we propose (Fig. 7.3 steps a–e) involves the photogenerated OH• radicals which can easily abstract a hydrogen atom from cyclohexane (step b) leading to the formation of a cyclohexyl radical. The very fast reaction of O_2 with Fe(II)(step c) in the presence of an alkyl radical leads to the formation of a Fe(III)-peroxoalkyl complex (step d) which, in non-polar solvents, is known to undergo an intramolecular decomposition, yielding rapidly and selectively the corresponding ketone and the starting Fe(III)TDCP-P(OH) complex.[24]

$$C_6H_{12} + O_2 + h\nu \xrightarrow{Fe(III)TDCPP(OH)} C_6H_{10}O + H_2O \quad (2)$$

When the polarity of the environment is increased by introducing CH_2Cl_2 as a cosolvent (60%v/v), a mixture of cyclohexanol and cyclohexanone is obtained (Table 7.1). One may explain the aforementioned results considering that, in a more polar medium, intermediate radicals will diffuse out from the solvent cage more easily, and that the Fe(III)-OOR intermediate will then have a different fate. The influence of the reaction environment on the mode of cleavage of iron–alkylperoxo complexes has been studied with many iron porphyrin biomimetic systems.[25,26] A polar environment and the availability of protons seem to favor a heterolytic cleavage of their O–O bonds with the formation of high valent iron–oxo complexes (Fig. 7.3 step f). These species are equivalent to the iron–oxo intermediates of the catalytic cycle of cytochrome P450, where they lead to the hydroxylation of C–H bonds, giving the starting iron(III) porphyrin complex (steps g, h). Thus, the oxidation of cyclohexane may follow two supplementary routes : (i) a free-radical dependent autoxidation leading to both cyclohexanol and cyclohexanone, and (ii) a selective hydroxylation giving exclusively cyclohexanol and regenerating the starting iron(III) porphyrin complex (steps g, h). Interestingly, irradiation of Fe(III)TDCPP(OH) in the presence of PBN, which is supposed to only trap free radicals in the bulk, such as R$^\bullet$, RO$^\bullet$, ROO$^\bullet$ leads to the selective formation of cyclohexanol (Table 7.1). This result indicates that route (ii) is the only one responsible for oxidation product formation under these conditions, and the overall stoichiometry, given in Equation (3), corresponds to a dioxygenase type reaction.

$$2C_6H_{12} + O_2 + h\nu \xrightarrow{Fe(III)TDCPP(OH)} 2C_6H_{11}OH \qquad (3)$$

7.1.3 Heterogeneous systems

Generally speaking, the use of heterogeneous media or organized systems is a suitable means of controlling the reaction environment in order to promote specific processes of interest in biomimetic catalysis by iron porphyrins.[27–29] As a development of this important field of research, we have recently investigated photocatalytic composite systems in which the porphyrin complexes are caged inside Nafion®, a commercial polyfluoro sulfonated membrane consisting of sulphonic groups connected to a polymeric structure of polytetrafluoroethylene.[30] We have chosen this support because (i) it is chemically inert in strong oxidizing media; (ii) it is totally transparent to the light of interest in the metal porphyrin photochemistry; (iii) it is expected to cage the monocationic porphyrin complexes inside its large anionic cavities where the SO_3^- groups are located; and (iv) dioxygen concentration inside Nafion is considerably higher than in organic solvents.[31]

The absorption of Fe(III)TDCPP inside a membrane of Nafion is relatively easy. The resin, after swelling in alcohol, is kept in a mixture of CH_2Cl_2 and alcohol (ROH = iPrOH, EtOH) containing the dissolved iron porphyrin. The Uv–vis spectra of the modified resin after swelling with ROH are exactly those of the starting iron porphyrin dissolved in acidified alcohol media, which can be ascribed to the monocationic species having an alcohol molecule bound to the axial position Fe(III)TDCPP(ROH).[32,33] Therefore, we must conclude that heterogenization does not affect the nature of the iron porphyrin axial ligand. An ESR spin-trapping investigation indicates that heterogenization does not affect even the primary photochemical process which consists of a homolytic cleavage of the Fe(III) – axial ligand bond (Equation 4). In fact, photoirradiation of modified membranes in the presence of the spin trap PBN, inside the cavity of an ESR spectrometer, yields a spectrum exhibiting the typical signals of the adducts between PBN and alkoxy radicals.[34]

The very fast reaction of the photogenerated ferrous porphyrin with O_2 is expected to restore the starting Fe(III) complex (Fig. 7.1a), so inducing the reductive activation of O_2 itself and the subsequent oxidation of organic substrates as for Fe(III)TDCPP(OH) in homogeneous solution. In spite of the similarities between the primary photoprocesses of Fe(III)TDCPP(OH) and Fe(III)TDCPP(ROH), prolonged photoirradiation (6 h, λ = 300–400 nm) of the latter complex inside a Nafion membrane immersed in ROH containing cyclohexane (25% v/v) failed to give cyclohexanol or cyclohexanone in detectable amounts. This behavior is only seemingly in contrast with that observed for Fe(III)TDCPP(OH) if one takes into account that in the second case the photogenerated radicals are known to be much more reactive that alkoxy radicals in extracting hydrogen from hydrocarbons.

On the other hand, alkenes are expected to capture efficiently the photogenerated alkoxy radicals through an allylic hydrogen abstraction process (Fig. 7.5).

$$\text{Fe(III)TDCPP(ROH)} + h\nu \rightarrow \text{Fe(II)TDCPP} + \text{R}\overset{\bullet}{\text{O}} + \text{H}^+ \quad (4)$$

$$\text{RO}^{\bullet} + \bigcirc\!\!\!= \longrightarrow \text{ROH} + \bigcirc\!\!\!\!\!\!{}^{\bullet}$$

Fig. 7.5 Capture of photogenerated alkoxy radicals by alkenes.

This hypothesis is confirmed by experiments in which the intensity of the ESR signal of the adduct between PBN and alkoxy radicals is followed as a function of irradiation time (Fig. 7.6). We observed that the signal intensity is unchanged in the presence of cyclohexane, while it is reduced by about 75% if cyclohexene is present. Therefore, photoirradiation of Fe(III)TDCPP(ROH) in the presence of cyclohexene should lead to the formation of a ferrous porphyrin cyclohexenyl radical pair. Reaction of O_2 with these species in the strong acidic environment of Nafion can give hydroxylating species as in the photocatalytic cycle reported in Fig. 7.3.

The nature of the products obtained upon irradiation are totally in agreement with the above statement. In fact, cyclohexane undergoes

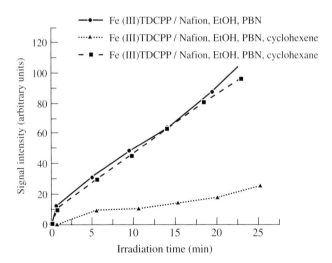

Fig. 7.6 ESR signal intensity of the adduct between PBN (3×10^{-2} mol dm^{-3}) and alkoxy radical versus irradiation time.

hydroxylation to cyclohexanol when cyclohexene is present as a cosubstrate (cyclohexane 12.5%, cyclohexene 12.5%). Moreover, cyclohexene is oxidized mainly to cyclohex-2-en-1-ol and *trans* cyclohexene-1, 2-diol monoethyl ether. The presence of this ether indirectly reveals the formation of cyclohexene epoxide which, in the strong acidic environment inside Nafion, undergoes a nucleophilic attack by the alcohol leading to ring opening. Interestingly, if the same experiment is carried out in the absence of cyclohexene the ether represents about 90% of the overall reacted alkene which is oxidized with a total turnover of about 1000.

The effect of heterogenization may be explained by assuming that the polymeric matrix favors the reactions occurring in proximity to the metal center while inhibiting the escaping of radical intermediates. The results described above show that heterogenization of iron porphyrin with Nafion creates new, promising photocatalysts which can successfully oxygenate alkanes and alkenes with O_2, in mild conditions and without consumption of reducing agents.

7.2 Titanium dioxide

7.2.1 Photo-oxidation processes

Irradiation ($\lambda > 360$ nm) of powder suspensions of TiO_2 in the presence of cyclohexane results in the formation of cyclohexanol, cyclohexanone and CO_2 (Table 7.2). As already discussed in the first paragraph (Fig. 7.1b), photoexcitation of TiO_2 particle promotes electrons in the conduction band and leaves positive holes in the valence band. These are able to oxidize the OH groups of the surface yielding $OH^\bullet$ radicals which are the likely initiators of the hydrocarbon oxidation according to Equations 5–8 not involving O_2 or its reduction products.[35,36]

$$\text{Ti-OH} + h^+ \rightarrow \text{Ti-OH}^{\bullet+} \qquad (5)$$

$$C_6H_{12} + (OH^\bullet)_{ads} \rightarrow C_6H_{11}^\bullet + H_2O \qquad (6)$$

$$C_6H_{11}^\bullet + (OH^\bullet)_{ads} \rightarrow (C_6H_{11}OH)_{ads} \qquad (7)$$

$$(C_6H_{11}OH)_{ads} + 2OH^\bullet \rightarrow C_6H_{10}O + 2H_2O \qquad (8)$$

Typical reactions that involve molecular oxygen and hydrocarbon radicals are the following:

$$C_6H_{11}^{\bullet} + O_2 \rightarrow C_6H_{11}OO^{\bullet} \tag{9}$$

$$C_6H_{11}OO^{\bullet} + C_6H_{12} \rightarrow C_6H_{11}^{\bullet} + C_6H_{11}OOH \tag{10}$$

$$C_6H_{11}OO^{\bullet} + e^-(Ti^{III}) \rightarrow C_6H_{10}O + Ti^{IV}OH \tag{11}$$

where the species are involved in adsorption–desorption equilibria. In addition, the alkyl radicals can react with the products of O_2 reduction according to the following Equations:

$$C_6H_{11}^{\bullet} + O_2^- \rightarrow C_6H_{10}O + OH^- \tag{12}$$

$$C_6H_{11}^{\bullet} + HO_2^{\bullet} \rightarrow C_6H_{10}O + H_2O \tag{13}$$

The reduction of peroxides (Equations 14, 15) is a source of radicals that leads to alcohol formation through reactions 5–8.

$$H_2O_2 + e^-(Ti^{III}) \rightarrow OH^{\bullet} + Ti^{IV} - OH \tag{14}$$

$$C_6H_{11}OOH + e^-(Ti^{III}) \rightarrow C_6H_{11}O^{\bullet} + Ti^{IV} - OH \tag{15}$$

It is a fact that selectivity is a key issue in the catalysis of chemicals production. Having this goal in mind, the use of TiO_2 presents a serious inconvenience in that, due to the high positive energy of the photogenerated holes, most organic compounds undergo complete oxidation to CO_2. This makes it an interesting photocatalyst for the detoxification of water as demonstrated by the large literature available on this subject.[37,38] Clearly, the exploitation of the stability and practical use of TiO_2 for photosynthetic purposes requires some strategy in the control of its high oxidation power.

7.2.2 Effect of the reaction medium

It is noteworthy that photoexcitation of TiO_2, as in the case of iron porphyrins, induces the formation of $OH^{\bullet}$ radicals which are the primary active species in the hydrocarbon oxidation. We were intrigued by the possibility of additional similarities between the homogeneous and the heterogenous photocatalytic systems. In particular, we were stimulated to investigate the effect of the polarity of the reaction medium on the photocatalytic properties of TiO_2, in accordance with the observation that the nature of the solvent has a remarkable effect on the selectivity of the photo-oxidation processes as described in the previous section.[39]

Table 7.2 shows the effect of the reaction medium on the photocatalytic oxidation of cyclohexane on TiO_2 suspensions. It reports the concentration of the oxygenation products and of CO_2 after irradiation of powder suspensions of TiO_2 in C_6H_{12} and its mixture with CH_2Cl_2. We show that, in the latter case, the yield in alcohol and ketone is higher, while that in CO_2 decreases. Therefore, we can claim an increase of the selectivity of the process since CO_2 is an undesired product. We propose that the main factor affecting the cyclohexanol to cyclohexanone concentration ratio is the chemisorption of products on the semiconductor surface.[40–42] We have experimental evidence that, when the composition of the medium is changed from pure cyclohexane to cyclohexane and CH_2Cl_2, the cyclohexanol adsorption on the semiconductor decreases, leading to accumulation of cyclohexanol in the liquid bulk. This can possibly account for the observed increase of the alcohol to ketone ratio increasing the amount of CH_2Cl_2 (Table 7.2).

7.3 Polyoxotungstates

7.3.1 $W_{10}O_{32}^{4-}$

The photochemical excitation of $(nBu_4N)_4W_{10}O_{32}$ ($W_{10}O_{32}^{4-}$) was carried out in a mixed solvent $CH_2Cl_2/C_6H_{12}/CH_3CN$ (6 : 3 : 1) at $\lambda = 325$ nm where there is an absorption maximum of the polyoxotungstate.[43] In agreement with previous work,[44–46] irradiation under an O_2 pressure of 760 torr leads to the oxidation of the alkane to cyclohexanol and cyclohexanone, without any irreversible modification of the photocatalyst (Table 7.3). In comparison with TiO_2, an interesting result is that the degradation of cyclohexane to CO_2 does not occur at all.

In order to gain more insight into the role of radical intermediates in the photocatalytic oxidation of cyclohexane, aerated solutions containing $W_{10}O_{32}^{4-}$ and C_6H_{12} were irradiated inside the ESR cavity in the presence of PBN. The spectrum obtained under irradiation consists of the superimposition of signals of different paramagnetic species. A rather intense triplet of doublets with hyperfine splitting constants ($a_N = 14.5$ G, $a_H = 2.1$ G) is consistent with the trapping of cyclohexyl radical.[47,48] The second signal observed, which appears only after several minutes of

Table 7.2 Photoassisted oxidation of cyclohexane on TiO_2 powder suspensions in different reaction medium after 90 min of irradiation ($\lambda > 360$ nm).

Sample	$[C_6H_{10}O + C_6H_{11}OH]$ (mol dm^{-3} × 10^3)	$C_6H_{11}OH$ /$C_6H_{10}O$	$[CO_2]$ (mol dm^{-3} × 10^4)	CO_2 % of the products
TiO_2 in neat C_6H_{12}	7.15	0.01	3.7	5.2
TiO_2 in C_6H_{12}/CH_2Cl_2 80/20 (v/v)	20.30	0.36	5.0	2.5
TiO_2 in C_6H_{12}/CH_2Cl_2 50/50 (v/v)	27.20	0.46	2.7	1.2

Table 7.3 Photocatalytic properties[a] of $W_{10}O_{32}^{4-}$ and of $W_{10}O_{32}^{4-}$ / Fe(III)TDCPP(OH) integrated system[b] in $CH_2Cl_2/C_6H_{12}/CH_3CN$ 6/3/1 mixed solvent, in the presence of 760 torr of O_2.

Photocatalysts	O_2 pressure	Φ_{ox}^c	$C_6H_{11}OH/C_6H_{10}O$	Recycle number[d]
$W_{10}O_{32}^{4-}$	760	0.35 ± 0.05	0.54	—
	20	0.28 ± 0.05	6.20	—
$W_{10}O_{32}^{4-}$ / Fe(III)TDCPP	760	0.30 ± 0.05	1.00	1200

a) Irradiations were carried out at 22 ± 1 °C, at 325 nm;
b) initial concentration of $W_{10}O_{32}^{4-}$: 2×10^{-4} mol dm^{-3}, initial concentration of Fe(III)Por: 3×10^{-5} mol dm^{-3};
c) ratio between the moles of oxidized cyclohexane and the moles of absorbed photons;
d) moles of oxidized cyclohexane per mole of consumed iron porphyrin.

irradiation, is weaker than the first one and consists of a triplet of doublets which can be attributed to the adduct between OH• radicals and PBN (a_N = 15.5 G, a_H = 2.8 G).[49]

These results show that irradiation produces $C_6H_{11}^{\bullet}$ radicals in the primary process according to equation 16. The detection of trapped OH• radicals is a likely consequence of the one electron reduction of H_2O_2 which has been proposed to be a possible intermediate in O_2 reduction by the photogenerated $W_{10}O_{32}^{5-}$.[50] The appearance of OH• radicals after an induction time can probably be explained by the fact that H_2O_2 is the main source of OH• radicals and must accumulate before any significant amount of them is detected.

$$C_6H_{12} + W_{10}O_{32}^{4-} + h\nu \rightarrow W_{10}O_{32}^{5-} + C_6H_{11}^{\bullet} \qquad (16)$$

The formation of both OH• and $C_6H_{11}^{\bullet}$ intermediates is expected to play a fundamental role in the subsequent formation of cyclohexanol and cyclohexanone as final products. In particular it may be attributed to (i) radical chain autoxidation processes and (ii) polyoxotungstate mediated decomposition of the $C_6H_{11}OOH$ species.

A strong influence of oxygen pressure on the product concentration ratio is shown in Table 7.3. In fact photoexcitation of $W_{10}O_{32}^{4-}$ in the presence of 760 torr of O_2 gives mainly cyclohexanone; conversely, the cyclohexanol to cyclohexanone ratio increases significantly by decreasing dioxygen concentration. The influence of O_2 concentration on cyclohexane product distribution may probably be explained by considering that, in the presence of small amounts of O_2, the reaction between bulk phase radical intermediates and O_2 are inhibited. Thus the concentration of $C_6H_{11}OOH$ decreases and, consequently, the probability of disproportionation reactions leading to cyclohexanone is reduced.

7.3.2 $W_{10}O_{32}^{4-}$/Fe^{III}porphyrin composite systems

We have also investigated the photocatalytic oxygenation of cyclohexane on composite systems which are able to combine the biomimetic properties of iron porphyrins with the high photochemical activity of polyoxotungstates.[51] Irradiation of $W_{10}O_{32}^{4-}$ in the presence of an iron porphyrin under an oxygen pressure of 760 torr leads to the oxidation of the substrate as summarized in Table 7.3. The presence of the iron porphyrin complex significantly affects the product distribution. In fact, photochemical excitation of the composite system $W_{10}O_{32}^{4-}/Fe^{III}TDCPP$ results in the formation of equal amounts of cyclohexanol and cyclohexanone, while photoexcitation of $W_{10}O_{32}^{4-}$ alone gives mainly cyclohexanone, as already discussed in the previous section. In this case, we assume that a biomimetic hydroxylating mechanism is operative, with the iron porphyrin complex reacting both in its ferrous and ferric forms. In particular, the reaction between Fe(III)Por and peroxides is confirmed by ESR spin trapping experiments, because we do not find the formation of OH• radicals even after an induction time. It is noteworthy that in comparison with iron porphyrins alone, higher yields in the oxidation products (about ten times) are obtained and the ratio between the moles of photo-oxidized cyclohexane and the moles of degraded porphyrin are significantly higher (about hundred times).

7.4 Conclusions

All the inorganic systems described in this article are efficient photocatalysts for the oxygenation of cyclohexane with molecular oxygen at room temperature and atmospheric pressure. They feature the following common characteristics: i) they use light as a 'clean' reagent to induce metal centered redox processes; ii) the function of O_2 is not just that of acting as an acceptor of electron from the photoreduced metal center but its reductive activation plays a fundamental role in the overall photoprocess; iii) the selectivity of the oxidation process can be controlled by changing the reaction environment.

Fe(III) porphyrins protected by sterically hindered substituents can be successfully used to oxidize cyclohexane to cyclohexanone or cyclohexanol through the formation of species which are key intermediates in enzymatic reactions. The nature of the dispersing medium is an important parameter in the control of the photocatalytic properties of TiO_2. In particular, by affecting the adsorption of intermediates, it is possible to address the photoprocess toward fine chemicals production. Polyoxotungstates are able to oxidize cyclohexane to cyclohexanol and cyclohexanone without degradation of the substrate to CO_2. In this case, the control of the alcohol to ketone concentration ratio has been achieved both by controlling O_2 concentration and using Fe(III) porphyrins/polyoxotungstate composite photocatalysts.

References

1. Suslick, K.S.; Watson, R.A. *Inorg. Chem.*, **1991**, *30*, 912–919.
2. Suslick, K.S.; Watson, R.A., *New J. Chem.*, **1992**, *16*, 633–642.
3. Bartocci, C.; Scandola, F.; Ferri, A.; Carassiti, V. *J. Am. Chem. Soc.*, **1980**, *102*, 7067–7072.
4. Bartocci, C.; Maldotti, A.: Carassiti, V.; Traverso, O.; Ferri, A. *Inorg. Chim. Acta.* **1985**, *107*, 5–12.
5. Peterson, M.W.; Rivers, D.S.; Richman, R.M. *J. Am. Chem. Soc.*, **1985**, *107*, 2907–2915.
6. Maldotti, A.; Amadelli, R.; Bartocci, C.; Carassiti, V.; Polo, E.; Varani, G. *Coord. Chem. Rev.* **1993**, *125*, 143–154.
7. Bard. A.J., *J. Photochem.*, **1994**, *10*, 59–63.
8. Maldotti, A.; Amadelli, R.; Bartocci, C. *J. Photochem. Photobiol., A: Chem.*, **1990**, *53*, 263–271.
9. Mathews, R.W. *Environ. Science Tech.*, **1991**, *25*, 460–465.
10. Giannotti, C.; Le Greneur, S.; Watts, O. *Tetrahedron Lett.*, **1983**, *24*, 5071–5072.
11. Mu, W.; Herrmann, J.M.; Pichat, P. *Catal. Lett.*, **1989**, *3*, 73–74.
12. Gerischer, H.; Heller, A. *J. Electrochem. Soc.*, **1992**, *139*, 113–118.
13. Linsebigler, A.; Lu, G.; Yates, J.T. Jr. *Chem. Rev.*, **1995**, *95*, 735–758.
14. Pope, M.T. *Heteropoly and Isopoly Oxometalates*, **1993**, Springer-Verlag.
15. Hill, C.L.; Bouchard, D.A.; Kadkhodayan, M.; Williamson, M.M.; Schmidt, J.A.; Hilinski, E.F. *J. Am. Chem. Soc.*, **1988**, *110*, 5471–5479.
16. Hiskia, A.; Papaconstantinou, E. *Inorg. Chem.*, **1992**, *31*, 163–167.
17. Meunier, B. *Chem. Rev.*, **1992**, *92*, 1411–1456.
18. Ellis, P.E.; Lyons, J.E. *Coord. Chem. Rev.*, **1990**, *105*, 181–193.
19. Hendrickson, D.N.; Kinnaird, M.G.; Suslick, K.S. *J. Am. Chem. Soc.*, **1987**, *109*, 1243–1244.
20. Weber, L.; Hommel, R.; Behling, J.; Haufe, G.; Hennig, H. *J. Am. Chem. Soc.*, **1994**, *116*, 2400–2408.
21. Maldotti, A.; Bartocci, C.; Amadelli, R.; Polo, E.; Battioni, P. Mansuy, D. *J. Chem. Soc. Chem. Comm.*, **1991**, 1487–1489.

22. Maldotti, A.; Bartocci, C.; Varani, G.; Molinari, A.; Battioni, P.; Mansuy, D. *Inorg. Chem.*, **1996**, *35*, 1126–1131.
23. Bartocci, C.; Maldotti, A.; Varani, G.; Battioni, P.; Carassiti, V.; Mansuy, D. *Inorg. Chem.*, **1991**, *30*, 1255–1259.
24. Cheng, R.L.; Latos-Grazynsky, L.; Balch, A.L. *Inorg. Chem.*, **1982**, *21*, 2412–2418.
25. Traylor, T.G.; Xu, F. *J. Am. Chem. Soc.*, **1990**, *112*, 178–186.
26. Gopinath, E.; Bruice, T.C. *J. Am. Chem. Soc.*, **1991**, *113*, 4657–4665.
27. Barloy, L.; Battioni, P.; Mansuy, D. *Chem. Comm.*, **1990**, 1365–1367.
28. Stiefel, E.I. In *Bioinorganic Catalysis*, **1993**, pp. 21–27 (ed. Reedijk, J.), Dekker, M. Inc., New York, Basel, Hong Kong.
29. Ramamurthy, V. *Photochemistry in Organized and Constrained Media*, **1991** (ed. Ramamurthy, V.). VCH, New York, Weinheim, Cambridge.
30. Maldotti, A.; Molinari, A.; Andreotti, A.; Fogagnolo, M.; Amadelli, R. *Chem. Comm.*, **1998**, 507–508.
31. Ogumi, Z.; Kuroe, T.; Takehara, Z. *J. Electrochem. Soc.*, **1985**, *132*, 2601–2605.
32. Bizet, C.; Morliere, P.; Brault, D.; Delgado, O,; Bazin, M.; Santus, R. *Photochem. Photobiol.*, **1981**, *34*, 315–320.
33. Maldotti, A.; Bartocci, C.; Amadelli, R.; Carassiti, V. *J. Chem. Soc. Dalton Trans.*, **1989**, 1197–1201.
34. Ledwith, A.; Russell, P.J.; Sutcliffe, L.M. *Proceed. Royal Soc. London, A*, **1973**, *332*, 151–154.
35. Gerischer, H. *Electrochim. Acta*, **1993**, *38*, 3–9.
36. Gerischer, H.; Heller, A. *J. Phys. Chem.*, **1991**, 95 5261–5267.
37. Hoffmann, M.R.; Martin, S.T.; Choi, W.; Bahnemann, D.W. *Chem. Rev.*, **1995**, *95*, 69–96.
38. Bahnemann, D.W.; Cunningham, J.; Fox, M.A.; Pichat, P.; Pelizzetti, E.; Serpone, N. In *Aquatic and Surface Photochemistry*, (ed. Helz G.R., Crosby D.G., Zepp R.G.), **1994**, pp. 261–279, CRC.
39. Boarini, P.; Carassiti, V.; Maldotti, A.; Amadelli, R. *Langmuir*, **1998**, *14*, 2080–2085.
40. Whalen, J.W. *J. Phys. Chem.*, **1962**, *66*, 511–515.
41. Fox, M.A.; Ogawa, H.; Pichat, P. *J. Org. Chem*, **1989**, *54*, 3847–3852.
42. Gilbert, L.; Mercier, C. In *Heterogeneous Catalysis and Fine chemicals III* (eds Guisnet, M., Barbier, J., Barrault, J., Bouchoule, C., Duprez, D., Perot G., Montassier, C.), **1993**, pp. 51–66. Elsevier, The Netherlands.

43. Maldotti, A.; Amadelli, R.; Carassiti, V.; Molinari, A. *Inorg. Chim. Acta*, **1997**, *256*, 309–312.
44. Hill, C.L.; Renneke, R.F.; Combs, L. *Tetrahedron*, **1988**, *44*, 7499–7507.
45. Renneke, R.F.; Hill, C.L. *J. Am Chem. Soc.*, **1988**, *110*, 5461–5470.
46. Ermolenko, L.; Giannotti, C. *J. Mol. Catal.*, **1996**, *114*, 87–91.
47. Janzen, E.G.; Hair, D.L. In *Advances in Free Radical Chemistry*, (ed. Tanner, D.D.) **1990**, Jau Press, London.
48. Iwahashi, H.; Ishikara, Y.; Sato, S.; Koiano, K. *Bull. Chem. Soc. Jpn.*, **1977**, *50*, 1278–1282.
49. Harbour, J.R.; Chow, V.; Bolton, J.R. *Can. J. Chem.*, **1974**, *52*, 3549–3555.
50. Lu, G.; Linsebigler, A.; Yates, J.T. Jr. *J. Phys. Chem.*, **1995**, *99*, 7626–7631.
51. Maldotti, A.; Molinari, A.; Bergamini, P.; Amadelli, R.; Battioni, P.; Mansuy, D. *J. Mol. Catal. A: Chemical*, **1996**, *113*, 147–157.

8 Catalytic hydrocarbons oxidation under fluorous/organic two-phase conditions

Fernando Montanari, Gianluca Pozzi, and Silvio Quici

8.1 Introduction

Catalysis plays a central role in the development of environmentally safe and clean chemical processes. Efficient, selective organic reactions that are otherwise unfeasible can be carried out under mild conditions in the presence of the proper catalyst, thus avoiding the use of hazardous reagents and drastically reducing the outflow of potentially polluting by-products.[1,2] Homogeneous and heterogeneous catalysis are often regarded as two distinct options, the latter being preferable for industrial applications. Indeed, both have their own advantages and limitations (Fig. 8.1).[3]

Homogeneous processes offer the striking possibility of tailoring the catalysts at the molecular level, since the active species may be characterized together with the catalytic cycle. Deactivation pathways can also be identified and, when possible, avoided. This deep understanding of the catalytic system leaves room for further improvement in its performance. The main problem with homogeneous catalysis is the separation of products and starting reagents from the catalyst and the recycling of the latter, usually an organometallic compound, which can be very expensive on account of its elaborate and possibly low yielding synthesis. A quantitative recovery and recycling of the catalyst is also necessary from the environmental point of view because transition metals are potentially toxic.

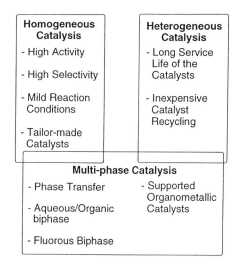

Fig. 8.1 Homogeneous versus heterogeneous catalysis.

In order to combine the advantages of homogeneous and heterogeneous catalysis, a number of multiphase catalytic systems have been studied. The best known example is 'phase transfer catalysis' (PTC) that was introduced about 30 years ago.[4] This technique often features an organic phase containing the substrate and an aqueous phase containing an alkali metal salt of an anionic reagent. The latter is transported into the organic phase by catalytic amounts of lipophilic ammonium or phosphonium cations located in the organic phase. PTC processes are widely used in anion promoted reaction and some of them have found application in industrial processes, because they allow reduced reaction times and mild reaction conditions.[5] Although this methodology also facilitates purification of the products, catalyst separation and recycling is still an open problem.

Other liquid–liquid two-phase systems in which the catalyst and the reaction products together with the residual starting materials, are located in different phases have been studied. In particular a number of laboratories, mostly in industry, have investigated aqueous–organic two-phase systems in which the organometallic catalyst resides in the aqueous phase, where the reaction takes place. A simple phase separation is then sufficient for isolating the products from the catalyst

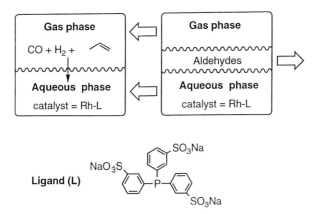

Fig. 8.2 Hydroformylation of propene in an aqueous/organic system (Ruhrchemie/Rhône-Poulenc).

which may be immediately recycled. This approach has been successfully applied to the hydroformylation of propene to butyraldehyde in an industrial process developed by Ruhrchemie/Rhône-Poulenc. The catalyst, a rhodium–phosphane complex, was made water soluble by the presence of sodium sulphonate substituents in the organic ligand. The whole process is schematically represented in Fig. 8.2.[6]

8.2 The 'fluorous biphase system' approach

Aqueous–organic two-phase systems are particularly appealing from the environmental point of view, but they suffer from a severe limitation since they only apply to water soluble substrates leading to water insoluble products. A new approach to liquid–liquid two-phase catalysis that overcomes this limitation has been proposed by Horvàth and Rabai.[7,8] Their 'fluorous biphase system' (FBS) basically consists of a perfluorinated or highly fluorinated solvent (the fluorous phase) and a second organic or inorganic solvent that is insoluble, or poorly soluble, in the fluorous phase. The term fluorous, coined in analogy with the more familiar term aqueous, can be applied to a spectrum of solvents. Many of them are commercially available and are generally regarded as non-toxic and biologically compatible, as proved by extensive experience with fluorocarbon coatings in cookware and artificial organ implants.[9] The use of perfluorinated solvents as reaction media has been described for a number of stoichiometric reactions.[10] In FBS catalysis, reagents and products are dissolved in the common organic solvent, while the catalyst is confined to the fluorous phase. At the end of the reaction a simple phase separation allows the recycling of the catalyst and the products recovery. The miscibility of perfluorinated solvents with organic solvents is temperature dependent so, in many cases, homogeneous solutions can be obtained at high temperature, whereas two immiscible phases are observed at room temperature. As a consequence homogeneous conditions can be maintained during the reaction step, thus having all the advantages of homogeneous catalysis, then lowering the temperature allows phase separation which is the main advantage of heterogeneous catalysis.

All these points have been clearly addressed by Horvàth and Ràbai in their study on the hydroformylation of lipophilic olefins (such as decene).[7] This reaction was performed using trifluoromethylperfluorocyclohexane ($CF_3C_6F_{11}$) as the fluorous phase and toluene as the second phase. The catalyst was prepared *in situ* from $Rh(CO)_2(acac)$ and $P[CH_2CH_2(CF_2)_5CF_3]_3$ (P/Rh = 40) under 75 psi of $CO/H_2 = 1:1$. The reaction was started by charging the autoclave with 1-decene at 100 °C under 150 psi of $CO/H_2 = 1:1$. Under these conditions the two-phase mixture becomes homogeneous and the reaction is almost complete in 11 h, affording undecanal in very high yield and n/iso selectivity. Phase separation was obtained by cooling the reaction mixture and the fluorous phase containing the catalyst was recycled showing unchanged activity. This process is shown in Fig. 8.3.

8.3 Oxidation of hydrocarbons under FBS conditions

For many years our research group has studied new oxidation catalysts based on metal complexes of tetraarylporphyrins, with the aim of performing selective hydrocarbon oxidations with cheap, safe and clean oxidants under very mild conditions.[11] Aqueous NaOCl at pH = 10.5, H_2O_2-30% at pH = 5.0 and CH_3COOOH/CH_3COOH in CH_3CN partially fulfil these requirements, but our secret dream was to find conditions allowing the activation of molecular oxygen. To this purpose,

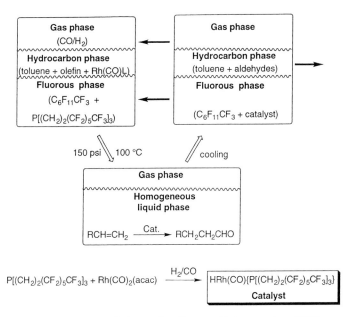

Fig. 8.3 Hydroformylation of alkenes in a fluorous biphase system (FBS).

in 1995, we started to investigate the possible use of metal complexes of perfluoroalkyl-substituted porphyrins. The presence of perfluoroalkyl chains (R_F) was expected to bring on selective solubility in fluorocarbons, which is the key feature for the use of such complexes as catalysts under FBS conditions. As soon recognized,[7,12] the chemical inertness of perfluorocarbons and the high oxygen solubility in these fluids would made the FBS approach particularly helpful for oxidation reactions. Due to the electron-withdrawing character of R_F substituents, increased stability of the porphyrin ligands under oxidizing conditions and increased catalytic activity through activation of the intermediate involved in the oxygen transfer to the substrate were also expected.[11] However, all these assumptions needed experimental confirmations, since only a few perfluoroalkyl-substituted porphyrins, among them compounds **1, 2, 3** and **4** (Fig. 8.4), had been reported at the time.[13,14,15,16] It is worth noting that **1–3** are not soluble in perfluorinated solvents and neither these porphyrins nor their metal complexes had been used as oxidation catalysts. On the other hand, compound **4** was

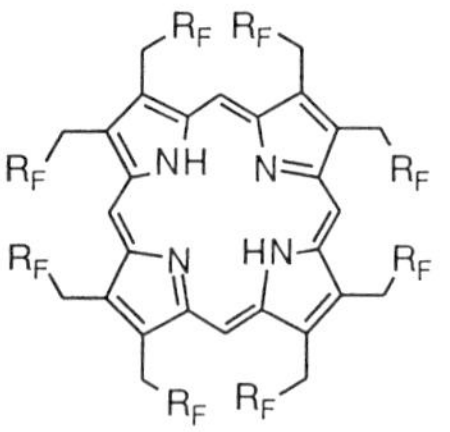

1

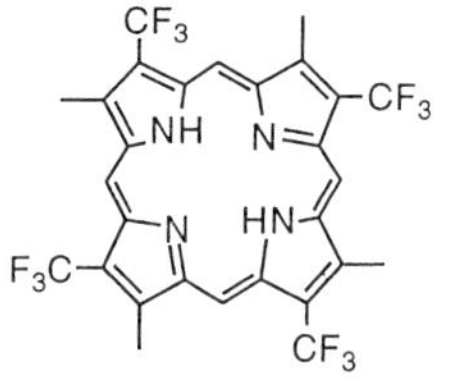

2

R_F = CF_3 or $CF_2CF_2CF_3$

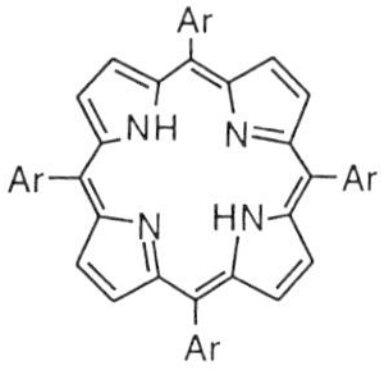

3

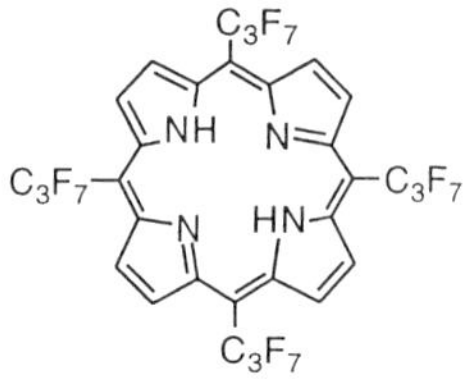

4

Ar = 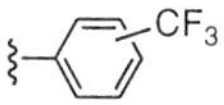

Fig. 8.4 Perfluoroalkyl-substituted porphyrins.

later used as fluorocarbon soluble sensitizer for the photo-oxidation of allylic alcohols to hydroperoxides under FBS conditions.[17]

Two synthetic pathways leading to perfluoroalkyl-substituted porphyrins can be, in principle, envisaged: i) functionalization of preformed porphyrin rings or ii) cyclization of perfluoroalkylated building blocks. In the first case, the meso aryl rings of the porphyrin must bear suitable functional groups allowing the covalent linkage of the RF residues. Thanks also to a long experience in the synthesis of porphyrins bearing several reactive functional groups,[11,18] this approach seemed to be simpler than the other, which involves the laborious synthesis of perfluorinated benzaldehydes. Nevertheless, we were able to synthesize a number of perfluoroalkyl functionalized porphyrins (Fig. 8.5) following the more appropriate pathway.

Porphyrins **5–8** have been prepared by condensation of the corresponding perfluoroalkyl aldehydes and pyrrole.[19] Equilibration of

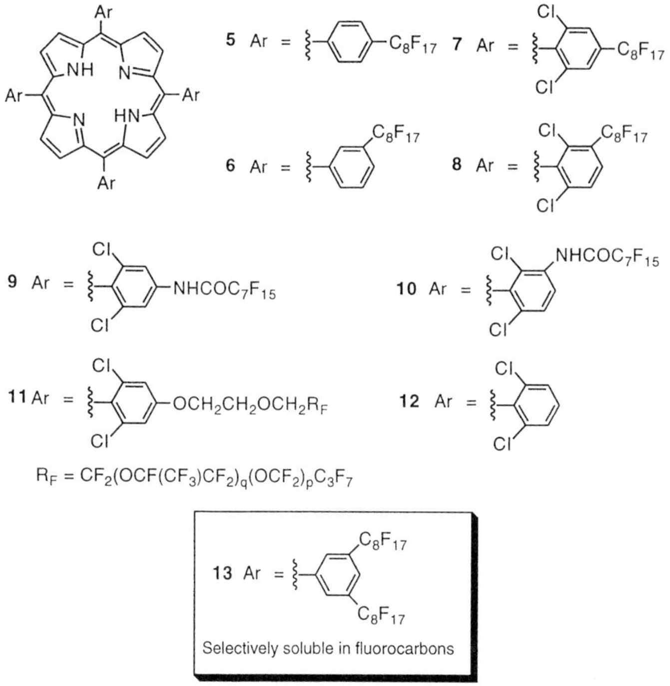

Fig. 8.5 New perfluoroalkyl-substituted tetraarylporphyrins.

10^{-2}M solution of equimolar amounts of aldehyde and pyrrole in CH_2Cl_2 in the presence of a Lewis acid (usually $BF_3 \cdot Et_2O$) and subsequent oxidation with 2,3-dichloro-5,6-dicyano-1,4-benzoquinone (DDQ) yielded the crude porphyrins that were purified by repeated column chromatography.[20] Porphyrins **9, 10** and **11** were prepared through acylation of amino- and alkylation of hydroxy-tetraarylporphyrins, respectively.[21]

The only porphyrin which was found to be selectively soluble in perfluorinated solvents was compound **13**, bearing two R_F tails on each meso aryl group.[22]

Mn(III)-complexes of **5–11** were nevertheless investigated as catalysts in alkene epoxydation carried out under classical aqueous/organic two-phase conditions by using NaOCl at pH = 10 as oxygen donor. Cyclooctene and 1-dodecene were used as models of reactive and poorly

Table 8.1 Epoxidation of Cyclooctene (S) by aqueous NaOCl (Ox) at pH 10.0 catalyzed by Mn(III)-complexes (P) of Porphyrins 5–10 in comparison with Mn(III)-12[a].

Entry	Catalyst	Time (min)	Yield[b] (%)	Selectivity[c] (%)
1	Mn-**5**	60[d]	8	95
2	Mn-**6**	60[d]	5	95
3	Mn-**7**	180	35	67
4	Mn-**8**	180	90	92
5	Mn-**9**	60[d]	38	73
6	Mn-**10**	60[d]	65	81
7	Mn-**12**	180	80	88

a Reaction temperature = 0 °C; Solvent = AcOEt; Molar ratio: S/P = 1000, L/P = 3, Ox/S = 2. L = N-hexylimidazole. b Determined by GC using the internal standard method. c Selectivity = (moles of epoxide)/(moles of substrate converted). d Yields does not improve on longer reaction times.

Table 8.2 Epoxidation of 1-dodecene (S) by aqueous NaOCl (Ox) at pH 10.0 catalyzed by Mn(III)-complexes (P) of porphyrins 7–10 in comparison with Mn(III)-12[a].

Entry	Catalyst	Time (min)	Yield[b] (%)	Selectivity[c] (%)
1[d]	Mn-**7**	180	/	/
2[e]	Mn-**8**	180	67	96
3[d]	Mn-**9**	240	22	65
4[d]	Mn-**10**	240	35	81
5[e]	Mn-**12**	240	33	69

a Reaction temperature = 0 °C; Molar ratio: S/P = 1000, L/P = 3, Ox/S = 2. L = N-hexylimidazole. b Determined by GC using the internal standard method. c Selectivity = (moles of epoxide)/(moles of substrate converted). d Solvent = AcOEt. e Solvent = CH_2Cl_2.

reactive alkenes, respectively, whereas the Mn(III) complex of tetrakis-(2,6-dichlorophenyl)porphyrin **12** was used as reference catalyst. Results are reported in Table 8.1 and Table 8.2 and can be summarized as follows:

i. the mere presence of R_F tails does not enhance the stability and catalytic activity of Mn-tetraarylporphyrins in the aqueous/organic biphasic system used; this experimental finding contrasts with what was suggested on the ground of computational studies;[23]
ii. the location of R_F tails and the steric protection provided by bulky ortho-substituents on the meso-phenyl rings are additional factors that effectively rule the performance of these catalysts;

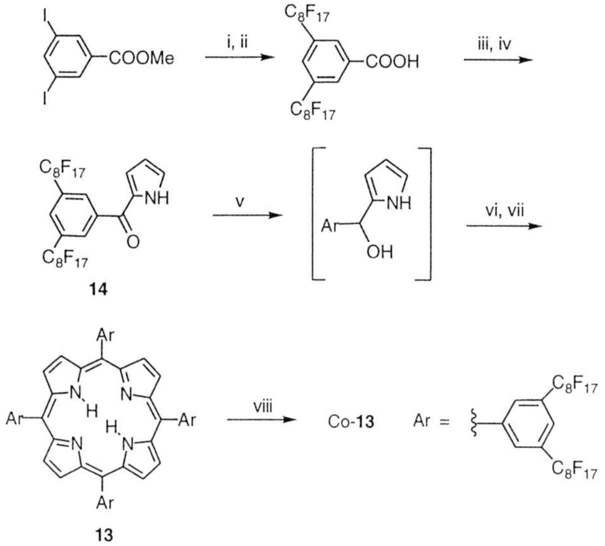

Fig. 8.6 Synthesis of the cobalt complex of porphyrin **13** bearing two perfluorooctyl chains on each meso aryl group.

i: n-C$_8$F$_{17}$I, Cu, DMF, T = 125 °C; *ii*: KOH, H$_2$O-MeOH, reflux; *iii*: SOCl$_2$, reflux; *iv*: Pyrrole + EtMgBr, Et$_2$O, 0 °C; *v*: LAH, THF; *vi*: Zn(OAc)$_2$·2H$_2$O, CH$_3$CH$_2$COOH, reflux, then 2,3-dichloro-5,6-dicyano-1,4-benzoquinone (DDQ); *vii*: CF$_3$COOH, reflux, then NaHCO$_3$ aq; *viii*: Co(OAc)$_2$·4H$_2$O, DMF, reflux.

iii. the presence of four R$_F$ tails has a clear beneficial influence in the epoxidation of poorly reactive linear α-olefins such as 1-dodecene.[20,21]

As already stated, only the cobalt complex of porphyrin **13** bearing two perfluorooctyl chains on each meso aryl group could be used as catalyst under FBS conditions. The synthesis of this compound is reported in Fig. 8.6.[22]

It should be pointed out that the aforementioned condensation of aldehyde and pyrrole,[19] failed in this case. Porphyrin **13** could be obtained through reduction of acylpyrrole **14** followed by tetramerization in propanoic acid at reflux in the presence of Zn(OAc)$_2$ · 2H$_2$O. The presence of Zn^{2+} as templating agent in the cyclization step turned out to

Table 8.3 Alkene (S) epoxidation by O_2/2-methylpropanal (A) catalyzed by Co-13 (P) in CH_3CN/perfluorohexane[a].

Entry	Substrate	Time (h)	Conversion[b] (%)	Selectivity[b,c] (%)
1	cis-Cyclooctene[d]	6	/	/
2	cis-Cyclooctene	3	100	> 95
3	Norbornene	5	95	> 95
4	1-Methylcyclohexene	3	100	> 95
5	2-Methylundec-1-ene	5	80	90
6	1-Dodecene[e]	14	60	87

a Molar ratio: S/P = 1000, A/S = 2. b Determined by GC using the internal standard method. c Selectivity = (moles of epoxide)/(moles of substrate converted). Epoxides were identified by comparison with authentic samples or by GC–MS. d Reaction carried out in the absence of Co-**13**. e A = 3-Methylbutanal.

be essential for obtaining the porphyrin in reasonable yield. After removal of the Zn cation, the free base porphyrin **13** was treated with $Co(OAc)_2 \cdot 4H_2O$ in DMF at reflux.[24]

The cobalt complex thus obtained proved to be an efficient catalyst for the epoxidation of alkenes by molecular oxygen and 2-methylpropanal as reducing agent. Reactions were carried out at room temperature under fluorous–organic two-phase conditions. A 10^{-4} M solution of Co-**13** in perfluorohexane was added to a solution of alkene in CH_3CN containing a twofold excess of aldehyde with respect to substrate. The resulting biphasic mixture was vigorously stirred under atmospheric pressure of dioxygen in a flask carefully shielded from direct sunlight. Under these reaction conditions, but in the absence of Co-**13**, alkenes did not react. The catalyst was completely partitioned in the fluorous phase as indicated by UV–Vis spectra of the two layers. At the end of the reaction, the two phases were easily separated and the CH_3CN layer was analyzed by gas-chromatography. The perfluorocarbon layer could be easily recovered and reused as such for a second run, still maintaining its activity. Results are reported in Table 8.3.

Conversions were generally very high for internal olefins while, as expected, the less reactive 1-alkenes (such as 1-dodecene) gave lower conversions. Selectivity was very high with any substrate. The beneficial effect of fluorous two-phase conditions is better appreciated by comparing these results with those reported in the literature: we were able to use a substrate/catalyst (S/C) ratio = 1000 in opposition to

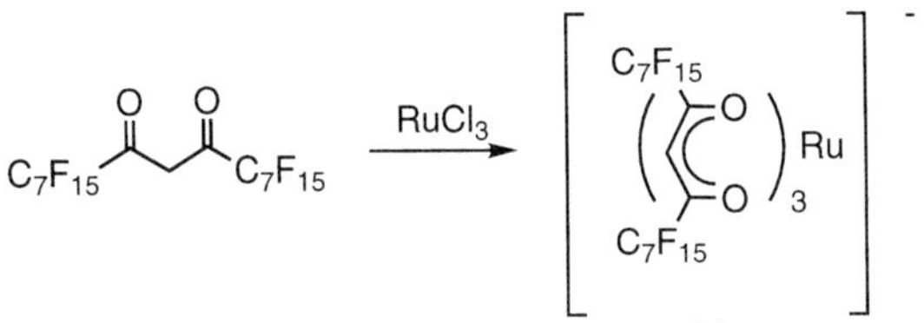

Fig. 8.7 Kmochel's catalyst

S/C = 20.[25] The only apparent drawback of this approach is that the preparation of the catalyst was complicated by the number of low-yielding synthetic steps required and by the tedious purification of the porphyrin. Knochel and coworkers also used a combination of O_2 and 2-methylpropanal as oxidizing agent for the catalytic epoxidation of internal olefins in the presence of 5% mol of ruthenium complex **14** (Fig. 8.7).[26] Reactions were carried out at 60 °C under homogeneous conditions. The solvents, toluene and $C_8F_{17}Br$, separated upon cooling and the catalyst confined in the fluorinated phase was easily recycled. Although **14** was ineffective in the epoxidation of 1-alkenes, two noteworthy features of this catalyst are the facile synthesis and the good selectivities obtained in the oxidation of polyfunctional substrates.

In order to match catalytic efficiency and synthetic simplicity, we prepared the polyazamacrocyclic ligand **15** in which the four R_F chains, ensuring solubility in the fluorous phase, were introduced by a straightforward, one-step N-alkylation (Fig. 8.8).[27] Commercially available 1,4,8,11-tetraazacyclotetradecane (cyclam) was chosen as starting compound, because of its low cost and easy handling. Furthermore cyclam and its alkyl derivatives form very stable complexes with many transition metal cations.[28] The (per)fluorooxyalkylenic alcohol **16** (Fig. 8.8), was supplied by Ausimont S.p.A. Milan.[29]

The presence of $CH_2OCH_2CH_2$ spacers between the binding sites and the R_F residues is crucial in maintaining the complexing ability of the macrocycle. It also prevents elimination during the N-alkylation step, which would be otherwise favored by the electron-withdrawing effect of the R_F residues. Ligand **15** is a thick oil, freely soluble in cold Et_2O and perfluorocarbons, but completely insoluble in cold hydrocarbons. When an excess of solid CuCl was added to a colorless 10^{-2} M solution of

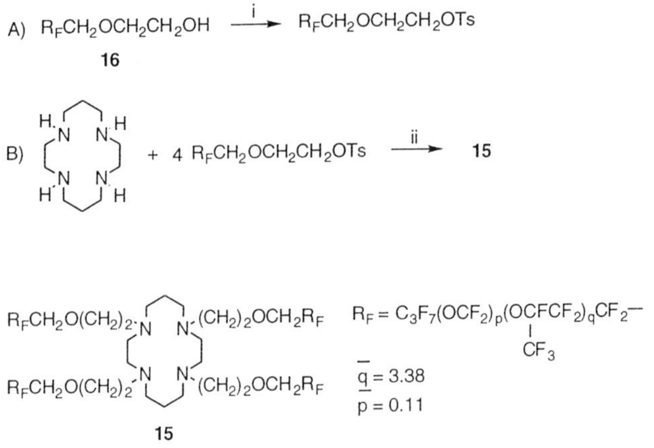

A) $R_FCH_2OCH_2CH_2OH \xrightarrow{i} R_FCH_2OCH_2CH_2OTs$
   16

B) [cyclam structure] + 4 $R_FCH_2OCH_2CH_2OTs \xrightarrow{ii}$ **15**

$R_FCH_2O(CH_2)_2$-N   N-$(CH_2)_2OCH_2R_F$ $R_F = C_3F_7(OCF_2)_p(OCFCF_2)_qCF_2-$
                                                                    |
$R_FCH_2O(CH_2)_2$-N   N-$(CH_2)_2OCH_2R_F$ CF_3
                                                 $\bar{q} = 3.38$
                15 $\bar{p} = 0.11$

i: p-Toluensulfonyl chloride, CH_2Cl_2/50% aqueous NaOH; *ii*: CH_3CN, Na_2CO_3, reflux 24 h.

Fig. 8.8 Synthesis of polyazamacrocyclic ligand **15**.

15 in perfluorohexane, the liquid phase turned immediately green, thus indicating copper complexation. This procedure failed for the preparation of Ni(II) and Co(II) complexes from $Ni(ClO_4)_2$ and $Co(ClO_4)_2$, respectively. This was not totally unexpected because, due to its low dielectric constant, perfluorohexane is unable to dissolve inorganic salts and cannot stabilize complexes with a strong ionic character. A Co(II) complex of ligand **15** was obtained in the presence of the fluorophilic $C_7F_{15}COO^-$ as counterion. $Co(C_7F_{15}COO)_2$, a pink solid only slightly soluble in fluorocarbons, was readily solubilized in perfluorohexane by adding one molar equivalent of **15** (Fig. 8.9).

The catalytic activity of the perfluorohexane solutions of the Cu(I) and Co(II) complex of **15** was tested in the FBS oxidation of cyclohexene and cyclooctane, in the presence of 80%-*t*-BuOOH in (*t*-BuO)$_2$O and molecular oxygen (Table 8.4). The hydrocarbon, acting both as the substrate and the organic phase, was used in large excess with respect to *t*-BuOOH. In the case of cyclohexene, copper and cobalt complexes gave similar results. The overall yields in cyclohexenone and cyclohexenol, obtained in 1 : 4 ratio, were higher than 100% with respect to *t*-BuOOH, thus indicating that O_2 was involved in the oxidation process. Cyclooctane was converted into a mixture of ketone

Catalytic hydrocarbons oxidation 157

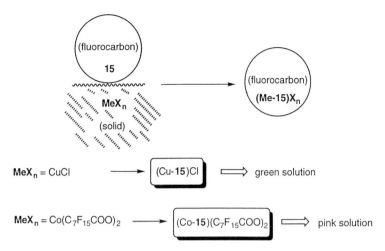

Fig. 8.9 Action of a perfluorohexane solution of ligand **15** on solid salts.

and alcohol in 1 : 4 ratio too, but only t-BuOOH was found to act as the oxidant. With both substrates, a slight decrease in catalytic activity was observed when a second run was carried out with the recycled fluorous phase.

A similar approach was described by Fish and coworkers who synthesized an N-perfluoroalkylated 1,4,7-triazacyclononane able to bind transition metals ions such as Co^{2+} and Mn^{2+}.[30] These complexes were used as catalysts in the oxidation of cyclohexene and cyclohexane, with results very close to those reported by us. Fish also presented

Table 8.4 Aerobic oxidation of hydrocarbons (S) catalyzed by metal complexes of ligand 15 (C) in the presence of t-BuOOH(O)[a].

Entry	Substrate	Metal	Selectivity[b] (%)		Overall yield[c,d] (%)	Turnover number[e]
			Alcohol	Ketone		
1	Cyclooctane	Cu	20	80	80	50
2	Cyclooctane	Co	24	76	30	19
3	Cyclohexene	Cu	25	75	514	320
4	Cyclohexene	Co	18	82	528	330

a Molar ratio: S/O = 80; O/C = 62.5. b Selectivity = (moles of product)/(moles of substrate converted). c Determined by GC using the internal standard method. d Based on the added t-BuOOH. e Turnover number = (mol of substrate converted/mol catalyst).

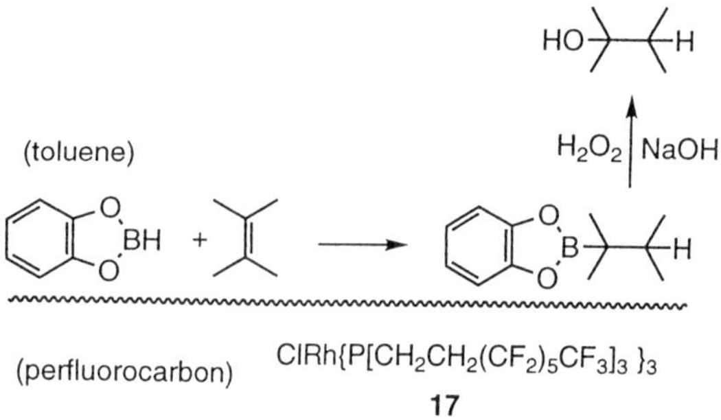

Fig. 8.10 Hydroboration–oxidation of alkenes.

evidences of an auto-oxydation mechanism involving alkoxy or alkylperoxy radicals.

Hydroboration of alkenes followed by oxidation is a well-known method for the synthesis of alcohols. An efficient catalytic FBS version of this process has been reported by Horvàth and Gladysz (Fig. 8.10).[31] The organic phase containing the alkene and catecholborane was stirred at T = 25–40 °C with a solution of the rhodium complex **17** in perfluoromethylcyclohexane. After 1–40 h the fluorous layer containing the catalyst was recovered and used again. Oxidative workup of the toluene solution of gave the alcohol in yields up to 90%.

8.4 Perspectives

Although the first results are very promising, much work remains to be done in order to widen the application of hydrocarbon oxidations under fluorous–organic two-phase conditions. The efforts of many groups are concentrated on the efficient synthesis of oxidation catalysts selectively soluble in fluorocarbons, which is fundamental to all cost-effective applications of FBS. For instance, our research is now oriented towards the development of enantioselective FBS catalysts, because the economic benefits of recovery and reuse of precious chiral complexes should reward the initial synthetic effort. We have already synthesized two optically active (salen)manganese(III) complexes, **18** and **19**, (Jacobsen–Katsuki catalysts) selectively soluble in fluorocarbons (Fig. 8.11).[32] They have been tested as catalysts in the epoxidation of alkenes under FBS conditions, in the presence of various oxygen donors.

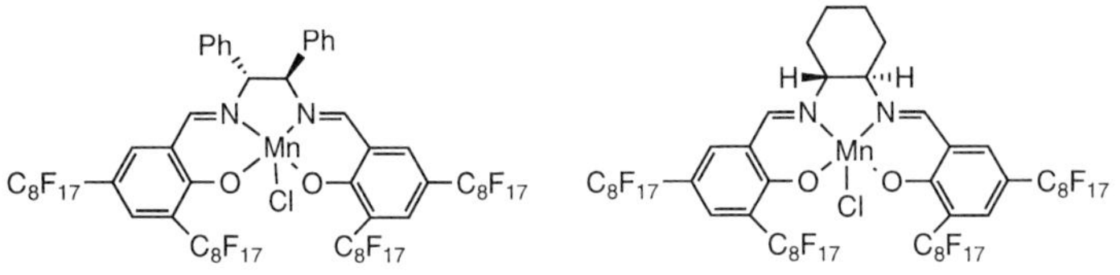

Fig. 8.11 Chiral Mn(III)-salen complexes

Distinct advantages were observed with respect to the classic homogeneous reactions: the catalysts were used in substantially lower amounts, their stability toward bleaching was increased and their activity remained high after the recycle. The easy separation of the products from the catalyst was another attractive feature of our system. Despite the good chemoselectivity and efficiency generally shown by our FBS catalysts, only indene was epoxidized with high enantioselectivy (>90%). All the other alkenes that we tested gave low ($\leq$15%) or even no enantiomeric excess. On the ground of these preliminary results, the rational synthesis of other chiral catalysts for FBS epoxidation of alkenes is now in progress.

References

1. Cusumano J.A. *Chemtech*, **1992**, *August*, 482–489 and references cited therein.
2. Simmons M.S. The role of catalysts in environmentally benign synthesis of Chemicals. In *Green Chemistry*; Anastas P.T., Williamson T.C. eds; *ACS Symposium Series* **1996**, *625*, 116–130.
3. Herrmann W.A., Cornils B. *Angew. Chem. Int. Ed. Engl.*, **1997**, *36*, 1049–1067 and references cited therein.
4. Montanari F., Quici S., Banfi S. Phase transfer catalysis. In *Comprehensive Supramolecular Chemistry—Supramolecular Technology. Vol. 10*; Reinhoudt D.N. ed.; Pergamon Elsevier Science; Oxford, **1996**, 389–416 and references cited therein.
5. Starks C.M., Liotta C.L., Halpern M. *Phase-Transfer Catalysis, Fundamentals, Applications and Industrial Perspectives*; Chapman and Hall; New York, **1994**.

6. Herrmann W.A., Kohlpaintner C.W. *Angew. Chem. Int. Ed. Engl.*, **1993**, *32*, 1524–1544 and references cited therein.
7. Horvàth I.T., Ràbai J. *Science*, **1994**, *266*, 72–75.
8. Horvàth I.T., Ràbai J. *European Patent Application* 0633062 A1, **1995**, Exxon Research and Engineering Company.
9. May G. *Chemistry in Britain*, **1997**, *August*, 34–36.
10. Zhu D.W. *Synthesis*, **1993**, 953–954.
11. Montanari F., Banfi S., Pozzi G., Quici S. Oxygenation reactions under two phase conditions. In *Metalloporphyrins Catalyzed Oxidation*; Montanari F., Casella L. eds; Kluwer Acad. Publ.; Dordrecht, **1994**.
12. Vogt M. *Zur Anwendung Perfluorierter Polyether bei der Immobilisierung Homogener Katalysatoren*, PhD. Thesis, **1991**, Rheinisch-Westfälischen Technischen Hochschule, Aachen.
13. Kaesler R.F., Le Goff E. *J. Org. Chem.*, **1982**, *47*, 5243–5246.
14. Ono N., Kawamura H., Maruyama K. *Bull. Chem. Soc. Japn.*, **1989**, *62*, 3386–3388.
15. Lindsey J.S., Wagner R. *J. Org. Chem.*, **1989**, *54*, 828–836.
16. Di Magno S.G., Williams R.A., Therien M.J. *J. Org. Chem.*, **1994**, *59*, 6943–6948.
17. Di Magno S.G., Dussault P.H., Schultz J.A. *J. Am. Chem. Soc.*, **1996**, *118*, 5312–5313.
18. Anelli P.L., Banfi S., Legramandi F., Montanari F., Pozzi G., Quici S. *J. Chem. Soc., Perkin Trans. 1*, **1993**, 1345–1357.
19. Lindsey J.S. The synthesis of meso-substituted porphyrins, in *Metalloporphyrins Catalyzed Oxidation*; Montanari F., Casella L. eds; Kluwer Acad. Publ.; Dordrecht, **1994**.
20. Pozzi G., Colombani I., Miglioli M., Montanari F., Quici S. *Tetrahedron*, **1997**, *53*, 6145–6162.
21. Pozzi G., Banfi S., Manfredi A., Montanari F., Quici S. *Tetrahedron*, **1996**, *52*, 11879–11888.
22. Pozzi G., Montanari F., Quici S. *Chem. Commun.* **1997**, 69–70.
23. Ghosh A. *J. Am. Chem. Soc.*, **1995**, *117*, 4691–4699.
24. Adler A.D., Longo F.P., Kampas F., Kim J. *J. Inorg. Nucl. Chem.*, **1970**, *32*, 2443–2445.
25. Mandal A.K., Khanna V., Iqbal J. *Tetrahedron Lett.*, **1997**, *37*, 3769–3772.
26. Klement I., Lütjens H., Knochel P. *Angew. Chem. Int. Ed. Engl.*, **1997**, *36*, 1454–1456.

27. Pozzi G., Cavazzini M., Quici S., Fontana S. *Tetrahedron Lett.*, **1997**, *38*, 7605–7608.
28. Roper J.R., Elias H. *Inorg. Chem.*, **1992**, *31*, 1202–1210 and references cited therein.
29. Sianesi D., Marchionni G., De Pasquale R.J. Perfluoropolyethers by perfluoroolefins photooxidation. In *Organofluorine Chemistry: Principles and Commercial Applications*; Banks R.E. ed.; Plenum Press; New York, **1994**.
30. Vincent J.-M., Rabion A., Yachandra V.K., Fish R.H. *Angew. Chem. Int. Ed. Engl.*, **1997**, *36*, 2346–2349.
31. Juliette J.J.J., Horvàth I.T., Gladysz J.A. *Angew. Chem. Int. Ed. Engl.*, **1997**, *36*, 1610–1612.
32. Pozzi G., Cinato F., Montanari F., Quici S. *Chem. Commun.*, **1998**, 877–878.

9 Strong acid solid materials as alternative catalysts in isobutane/2-butene alkylation

A. Corma and J.M. López Nieto

Abstract

The catalytic behavior of various strong acid solid catalysts, i.e. sulfated metal oxides, heteropolyacids and nafion–silica composites, for the isobutane/2-butene alkylation has been comparatively studied. The optimum reaction temperature depends on the acid strength which determines the cracking–alkylation–oligomerization ratio. In addition, catalyst decay is observed on all the catalysts studied. Finally, the nature of the active sites of these catalysts and the implications on the different reactions occurring and on the catalyst decay is also discussed.

9.1 Introduction

In an attempt to reduce emissions from automobile fuels, the composition of gasoline has been regulated in United States by the U.S. Environmental Protection Agency, starting in 1997.[1,2] As a results, important challenges in the oil refining industry are expected over the next years. These involve the reduction of the gasoline vapour pressure (RPV) and the content of alkenes, aromatics (specially benzene) and sulfur, with a complete elimination of lead additives.[1] Moreover, an increase in the minimum concentration of oxygenates in the gasoline pool will be mandatory.[2,3] A possible consequence of this is a decrease

Table 9.1 Characteristics of the main gasoline streams.

Stream	RVP	Aromatics	Olefins	RON	MON
FCC	7.1	29.2	29.1	92	81
Reformate	5.3	62.6	0.7	98	87
LRS naphta	10.0	9.8	2.2	79	76
Alkylate	7.9	0.4	0.5	93	91

in the total amount of gasoline available, together with a depletion of the gasoline octane number, unless some gasoline streams are further processed. In this scenario, it would be beneficial to have the largest possible contribution of alkylated gasoline to the pool of gasolines. Indeed, this stream presents high octane number and low sensitivity values, together with clean burning characteristics (Table 9.1).[1,3]

Isobutane can be combined with light olefins (C_2–C_5) in order to produce branched chained paraffinic fuel, especially trimethylpentanes (**TMP**) when C_4-olefins are used, which are the best motor octane components of a gasoline pool. Commercially, the gasoline alkylation process involves the use of the environmentally hazardous concentrated sulfuric or hydrofluoric acid, both operating at relatively low temperatures.[1,4] Table 9.2 presents the selectivity to the main reaction products during the isobutane/2-butene alkylation on inorganic acids. It can be seen that the selectivity to TMP is near to 90% in both cases. The results given in Table 9.3 indicate a worldwide alkylate capacity of about 1500 MBPCD is expected in 1998, and about 60% will use HF as catalyst. Since sulfuric acid is less hazardous than HF, one could expect a preferential increase in the number of alkylation plants with sulfuric

Table 9.2 Alkylation of isobutane with 2-butene on commercial liquid acid catalysts[a].

Catalysts	Selectivity (%) H_2SO_4[b]	HF[c]
C_5–C_7	7.5	2.7
C_8	88.1	93.2
C_{9+}	4.4	4.1
TMP/C_8	90.8	91.8
TMP/(DMH)	9.9	11.2

a At conversions higher than 98%, after ref. 1. b At 0 °C. c At 20 °C.

Table 9.3 World alkylation capacity[a].

Region	Capacity (MBPCD)	
	1995	1996
US	1057	1100
EEC	212	212
Asia Pacific	123	136
Total	1392	1448

a After Rao and Vatcha 1996.

acid as catalyst.[4] However, refiners using sulfuric acid alkylation units must ship large volumes of spent acid off site for regeneration, thus creating potential transportation hazards.[5] It has been proposed that the use of additives in HF alkylation units could be used to significantly reduce aerosol production.[6] Nevertheless, due to environmental pollution concerns and in order to solve the problems of corrosion associated with the use of H_2SO_4 and HF, there is a strong incentive for developing non-contaminant solid catalysts.

Obviously, these solids should present a strong surface acidity, be selective towards the formation of TMP and last for sufficient time. In the design of these catalysts, their acid strength is a parameter of paramount importance. Excessively strong acidities will drive the reaction towards cracking of the branched paraffins, while too low acidity will mainly produce oligomerization of the olefins. For instance, when the isobutane alkylation is carried out on a liquid acid catalysts, that is trifluoromethanesulfonic acid mixed with trifluoroacetic acid or water, the best alkylation conditions are obtained at an acid strength of about $H_0 = -10.7$ as determined by Hammet indicators.[7]

One way to reduce the negative impact of the H_2SO_4 is to use the acid impregnated on silica.[8] This not only decreases the hazards, but also increases the catalyst life. It has also been found that supported triflic acid, CF_3SO_3H, can act as a good alkylation catalyst.[9]

Lewis acids supported on metal oxides, specially $AlCl_3$[10,11] or BF_3[12] based catalysts have been proposed as alternative catalysts. However, the formation of HCl or HF indicates that this may not be a productive route.

Taking into account the possibility of tuning the acidity of zeolites, these have been widely used as acid catalyst.[13,14] In this field,

USY,[1,15,16] REY,[17] EMT and La-EMT[18,19] and Beta[16,20,21] zeolites have been proposed as selective catalysts in the alkylation of isobutane with butenes. Improvement of the catalyst characteristics have been demonstrated when a promoter such as BF_3 is introduced either on the catalyst or together with the feed.[22,23] The presence of BF_3 is particularly effective in the activation step. BF_3 toxicity, however, must also be considered.[23]

At the end of the 1970s, new solid materials with an hypothetical strong acid character were reported. Zirconium oxide treated with sulfate ion was identified as an active and selective catalyst in the isomerization of n-butane to isobutane at low temperature.[24,25] Other metal oxides such as TiO_2, SnO_2, Fe_2O_3, HfO_2, etc were susceptible to being converted in successful acid catalysts.[13,26,27,28] Extensive characterization studies have been published in the last years in which new aspects of the nature of active sites and the mechanism of the activation of short chain paraffins (including n-butane at room temperatures) have been proposed. However, the role and nature of the active sites on these SO_4^{2-} promoted oxides is still a matter of discussion.

Heteropoly acids (HPAs) and related compounds have attracted increasing interest in catalysis owing to their ability to catalyze both acidic and red–ox processes. In addition, they present strong acidities and can be prepared with high surface areas.[29,30] In fact, the pH values of aqueous solutions of heteropolyacids indicate that they are strong acids. This is related to the large size of the polyanion having a low and delocalized surface charge density, causing a weak interaction between polyanion and proton. In addition, they present strong acidic properties in the solid state and are very sensitive to countercations as well as to the constituent elements of polyanions. Thus, depending on their composition, heteropolyacids can be prepared in a wide range of acidities. It has been noted that heteropolyacids catalyze the isomerization of n-butane to isobutane at a reaction temperature higher than on zirconia sulfated catalysts.[31] This indicates that heteropoly compounds should provide acid sites which have a lower acid strength than sulfated metal oxides or that act through a different initiation mechanism.

Finally, acidic ion-exchange polymers represent a new type of solid acid material, although its acidity, or at least its capacity to activate paraffins, is lower than that of sulfated zirconia or even heteropolyacids. Among the resins Nafion and, in particular, perfluorosulfonic resins are of special interest.[32] They consist of a backbone of tetrafluoroethylene with pendent side chains of perfluorinated vinyl ethers which terminate in sulfonic acid groups. The general structure can be described as:

$$[-(CF_2-CF_2)_x CF-CF_2-]_y$$
$$[O-CF_2-CF(CF_2)_2]_z-O-CF_2-CF_2-SO_3H$$

($x = 5-13.5$; $y = 1000$; $z = 1, 2, 3, \ldots$)

Nafion resin has been proposed to be effective in a wide range of reactions catalyzed by acids.[33–35] The superacidity of the sulfonic group is attributed to the electron-withdrawing forces of the neighbouring $-CF_2CF_2SO_3H$ groups.[26] H-Nafion is commercially available although it provides a low surface area.

It has been proposed that high surface area Nafion resin/silica nanocomposites could be used as solid acid catalysts, and that they present a higher activity than the pure resin.[36] However, there can be some interactions between Nafion and silica.[37,38] In this way, it has recently been proposed that Nafion resin/silica nanocomposites with low surface area could present a better activity and selectivity than those with high surface area.[38]

In this paper, a comparative study of the catalytic behavior of the different solid acid catalysts mentioned above for the isobutane/2-butene alkylation is presented. In all solid catalysts studied, a catalyst decay is observed and therefore their intrinsic activity and selectivity will be calculated here at a short time on stream, generally at 1 mm of time on stream (TOS). We then also compared the corresponding catalyst decay by working at longer times on stream. In addition, the nature of the active sites on these catalysts and their implications in the different reactions occurring will also be described.

9.2 Reaction mechanism of isobutane/butenes alkylation

Before comparing the catalytic behavior of strong solids acids, it is useful to review the mechanism and the reactions occuring during the alkylation of isobutane with olefins.

It is generally accepted that the alkylation of isobutane with C_3–C_5 olefins is a complex process involving a series of simultaneous and

168 *Green Chemistry: Challenging Perspectives*

consecutive reactions occurring through carbocation intemediates.[1] The first step in this process is the formation of a carbenium anion by the addition of a proton to an olefin. This cation can react with isobutane forming a new cation (mainly t-butyl cation) or it can be added to an olefin forming the corresponding C_8-carbocation. The resultant carbocation intermediates can now:

1) desorb producing the corresponding trimethylpentanes (2,2,4- or 2,2,3-TMP) or dimethylhexanes (2,2-DMH);
2) or desorb, after an isomerization via hydride and methyl shift, forming more stable carbenium ions which are precursors of the trimethylpentanes (2,3,4-, 2,3,3-, or 2,2,4-TMP) observed in the products.

Besides the alkylation reaction, two competing reactions can also occur:

1) the cracking of alkylated products;
2) the oligomerization of the olefin feed.

In addition to this, and operating at high olefin conversions, consecutive reactions can also appear. This is the case in the cracking of large isoalkyl cations of hydrocarbons, especially by a β-scission route, forming C_5 to C_7 hydrocarbons. It must be noted that the acid strength of active sites contributes to enhance or suppress some of these reactions. In this sense,

Table 9.4 Influence of the reaction temperature on the catalytic behavior of SO_4^{2-}/ZrO_2 and H-Beta catalysts during the isobutane/2-butene alkylation[a].

	SO_4^{2-}/ZrO_2[b]		H-BETA[c]	
	50 °C	0 °C	50 °C	0 °C
2-butene conversion (%)	92.4	65.0	96.3	62.0
Distribution of C_{5+} (wt%)[b]				
C_5–C_7	65.9	19.5	24.9	2.7
C_8	27.8	53.6	45.3	69.4
C_{9+}	6.3	9.0	20.8	27.9
TMP/C_8	72.6	91.2	67.2	16.4
TMP/(DMH-$C_8^=$)	3.4	24.0	2.4	0.7

a) Catalytic results obtained at a TOS of 1 min. Reaction conditions: Isobutane/2-butene molar ratio of 15 and 25 bar total pressure.   b) WHSF, referred to 2-butene, of 1 h^{-1}.
c) WHSF, referred to 2-butene, of 2 h^{-1}.

the acidity required to carry out each of the above reactions decreases in the following order: cracking > alkylation > oligomerization. Therefore, the selectivity of a given catalysts towards one or another reaction will depend, in addition to other factors, on its acid strength distribution. Furthermore, the rate of desorption of the alkyl carbocation intermediates will depend on the hydrogen transfer ability, and this will contribute to enhance the selectivity to TMP products.

9.3 Isobutane alkylation on solid superacid catalysts

9.3.1 Alkylation reaction on sulfated metal oxides

Guo et al.[39] and Corma et al.[1,21,40–42] compared the activity of SO_4^{2-}/ZrO_2 and other oxide sulfate catalysts in the alkylation of isobutane with butenes at different reaction temperatures. While in the first case the product distribution obtained was not discussed,[39] it has been shown in the other reports that sulfated zirconia (SZ) catalysts show interesting catalytic performances during the alkylation of isobutane with olefins, and a good selectivity to TMP can be obtained by properly selecting the reaction conditions.[1,21,40]

Table 9.4 shows the catalytic behavior of a SZ catalyst at different reaction temperatures at 1 min times on stream (TOS).[21] For comparison, the results obtained on an active Beta have also been included. At high reaction temperature, that is 50 °C, the SZ catalyst produces a very high amount of cracked products (C_5–C_7) and it is less selective to alkylation (TMP) than zeolite Beta. However, a completely different result is observed at a lower reaction temperature, that is 0 °C, which is similar to the temperature used in industrial processes. Under these conditions, although SZ still shows some cracking activity, it produces little C_{9+} products and is highly selective towards the formation of C_8 hydrocarbons and TMP (specially the 2,2,4-trimethylpentane).

The different catalytic performance of Beta zeolite and sulfated zirconia with reaction temperature has been explained on the basis of the higher acid strength of the SZ catalysts.[41,42] These sites could be responsible for both alkylation and cracking reactions at high temperatures while they will mainly favor alkylation reactions at low reaction temperatures.[41]

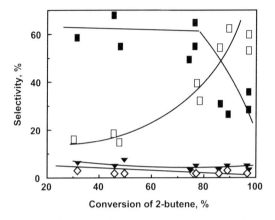

Fig. 9.1 Variation of the selectivity to the main reaction products with the conversion of 2-butene obtained on SZ catalysts at a 1 min TOS and 30 °C. Symbols: trimethylpentanes (■); C_5–C_7 (□); DMH (▼); C_8-olefins (◇). Reaction conditions: Isobutane/2-butene molar ratio of 15, 25 bar total pressure and a WSHF, referred to 2-butene, of 2 h^{-1}.

Since it has been proposed that the presence of sulfate ions and the method of preparation determine both the catalytic activity and the number of surface sulfate species,[41] it becomes of interest to compare the catalytic performance for isobutane/butene alkylation of catalysts with different sulfur-loading, and different catalyst preparation procedures. Figure 9.1 shows the variation of the selectivity to the main reaction products obtained at different levels of 2-butene conversion (achieved at a TOS of 1 min) on SZ catalysts prepared by different methods. It can be seen that the selectivity to the main reaction products strongly changes from one preparation to another. However, by comparison of these results with those obtained at different alkene space velocity (WHSV), on an active and selective catalyst, it can be seen that the selectivity to TMP obtained on SZ catalysts prepared by different methods is only a function of their catalytic activity (Fig. 9.2).[45] It is now clear that the catalyst preparation procedure can determine the number of the sulfur atoms on the surface and, consequently, the number of acid sites present. However, it must be noted that the nature of sulfate species and the strength of acid sites do not change with the catalyst preparation method.

It is clear that if the catalytic behavior of solid acids depends on their acid strength, the selectivity to alkylated products could be modified by

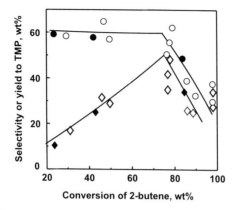

Fig. 9.2 Selectivity (●, ○) and yield to TMP (◆, ◇) obtained during the isobutane/2-butene alkylation on SZ catalysts at 30 °C: a) catalytic results obtained on catalysts with different S-content and/or prepared with different catalyst preparation procedures (○, ◇); b) catalytic results obtained, at different WSHF, on an active catalyst (●, ◆). Reaction conditions as Fig. 9.1.

tuning the acid nature of the support. In this way, the acid strength of the catalysts could be modified, in principle, by changing the anion incorporated or the metal oxide support. It has been reported that although ZrO_2 and TiO_2 can be modified with several acids, that is NH_4F, SF_4, $CHClF_2$, HCl, H_2SO_4 and HPO_3, only those treated with sulfuric acid are active and selective catalysts for the alkylation of isobutane.[43] On the other hand, strong differences can be obtained by changing the nature of the support,[44,45] as shown in Fig. 9.3 where the catalytic activity and the selectivity to TMP obtained in the isobutane/2-butene alkylation on sulfated metal oxides, that is ZrO_2 (SZ), TiO_2 (STi) and SnO_2 (SSn), have been presented. Both the selectivity to TMP and the TMP/DMH ratio at isoconversion conditions for SZ is higher than for STi and SSn. In this way, it has been observed that the activity in the conversion of 2-butene during the isobutane/2-butene alkylation decreases as follow: $SO_4^{2-}/ZrO_2 > SO_4^{2-}/TiO_2 > SO_4^{2-}/SnO_2 > SO_4^{2-}/Fe_2O_3 > SO_4^{2-}/Al_2O_3$. In the same way as in the case of SZ, the selectivity to TMP obtained on STi and SSn appears to be only a function of the number of acid sites (related to sulfate anions) which also determines the catalytic activity for the 2-butene conversion.

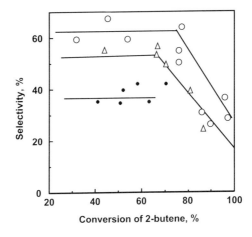

Fig. 9.3 Variation of the selectivity to TMP with the 2-butene conversion on SZ (○), STi (△) and SSn (●) catalysts at 30 °C and 1 min TOS. Reaction conditions as Fig. 9.1.

Finally, it is clear that, as was observed before in the case of SZ and Beta catalysts, the catalytic properties can be coupled with an adequate selection of the reaction temperature in order to achieve the higher selectivity to TMP on sulfated metal oxides. In this way, the higher the acid strength of the acid sites of the catalyst the lower the reaction temperature required.

9.3.2 Alkylation reaction on heteropolyacids catalysts

Heteropolyacid based catalysts have been recently studied for the isobutane/2-butene alkylation. This is the case of Cs^+, NH_4^+ and K^+ salts of 12-tungstophosphoric acids, $Me_xH_{3-x}PW_{12}O_{40}$, with different stoichiometries[46] or on different supports.[47] Table 9.5 shows the catalytic behavior, at 1 min TOS, of the most active and selective unsupported heteropoly compounds. It can be seen that the catalytic activity strongly depends on the number and type of the countercation incorporated, which determines the number and strength of the surface acid sites. In this way, the catalytic activity and the effectiveness to TMP decreases as follows: $K_{2.5}H_{0.5}PW_{12}O_{40}$ > $Cs_{2.5}H_{0.5}PW_{12}O_{40}$ > $H_3PW_{12}O_{40}$ > $(NH_4)_2HPW_{12}O_{40}$. However, a correlation between

Table 9.5 Catalytic properties in the isobutane/2-butene alkylation of acidic salts of 12-tungstophosphoric acids at 80 °C[a].

	H_3PW	$(NH_4)_2HPW$	$Cs_{2.5}H_{0.5}PW$	$K_{2.5}H_{0.5}PW$
2-butene conversion (wt%)	56.9	45.3	73.1	85.7
Distribution of C_{5+} (wt%)				
C_5–C_7	36.5	36.2	21.0	30.5
C_8	56.4	63.2	67.5	55.5
C_{9+}	7.1	6.8	11.5	3.4
TMP/C_8	62.8	54.4	57.0	74.0
TMP/(DMH + $C_8^=$)	1.7	1.2	1.3	2.85

a Catalytic results obtained at a TOS of 1 min. Reaction conditions: isobutane/2-butene molar ratio of 15 and 25 bar total pressure.

activity and the number and strength of acid sites has not been observed, and it has been suggested that the existence of mesoporosity could be more important than differences in the acid character of the catalysts.[48]

Heteropolyacids present, generally, a low surface area. However, this can be increased by supporting them on high surface area low reactivity inorganic solid carriers. It has been recently reported that SiO_2- or MCM-41-supported $H_3PW_{12}O_{40}$ are effective catalysts for the production of TMP during the isobutane/2-butene alkylation.[47] Table 9.6 shows comparatively the catalytic performance of pure and SiO_2-supported 12-tungstophosphoric acid (H_3PW/SiO_2; 20wt% of H_3PW) catalysts in the alkylation of isobutane at a reaction temperature of 33 °C and 1 min TOS. In the case of pure H_3WP, the selectivity to TMP increases when decreasing the reaction temperature (see Tables 9.5 and 9.6) while, in an opposite trend, the selectivity to cracked products decreases when increasing the reaction temperature.

When the 12-tungstophosphoric acid is supported on SiO_2 the catalytic activity strongly increases as a result of the higher dispersion of the active sites. In addition, a high selectivity and a high relative yield to TMP are also obtained.

Thus, it can be concluded that H_3WP based catalysts present strong acid sites which could operate at temperatures lower than zeolites but higher than SZ catalysts. In addition, the incorporation of H_3WP on

Table 9.6 Catalytic behavior of pure and SiO$_2$-supported 12-tungstophosphoric acid at 33 °C^a.

	H$_3$PW$_{12}$O$_{40}$	H$_3$PW(40wt%)/SiO$_2^b$
2-butene conversion (%)	27.2	98.8
Distribution of C$_{5+}$ (wt%)		
C$_5$–C$_7$	15.1	29.5
C$_8$	71.6	59.5
C$_{9+}$	13.3	11.0
TMP/C$_8$	55.4	85.3
TMP/(DMH-C$_8^-$)	1.24	5.81

a) Catalytic results obtained at a TOS of 1 min. Reaction conditions: isobutane/2-butene molar ratio of 15 and 25 bar total pressure. b) SiO$_2$-supported 12-tungstophosphoric acid.

metal oxides could offer the advantage of operating with relatively low amounts of active phase but with still the same catalytic performance.

9.3.3 Alkylation reaction on Nafion

Nafion resin has also been considered as a superacid material[7,49] and the catalytic behavior has been studied of a commercial Nafion-H in the isobutane/2-butene alkylation at 80 °C in a stirred semi-batch autoclave. In these studies, it was observed that at a butene conversion of 40% the selectivity to C$_8$ paraffins was 43.7, with a TMP/DMH molar ratio of 2.0 and an i-octenes/C$_8$ products weight ratio of 0.48. Similar results are obtained on a fixed bed reactor (Table 9.7). However, as observed in

Table 9.7 Catalytic properties in the isobutane/2-butene alkylation of Nafion and a nanocomposite of Nafion resin/silica catalysts at 80 °C^a.

	H-Nafion	H-Nafion/SiO$_2$
2-butene conversion (%)	41.1	95.0
Distribution of C$_{5+}$ (wt%)^b		
C$_5$–C$_7$	14.6	25.5
C$_8$	43.3	54.1
C$_{9+}$	42.0	20.4
TMP/C$_8$	42.0	51.1
TTMP/(DMH-C$_8^-$)	0.7	1.06

a) Catalytic results obtained at a TOS of 1 min. Reaction conditions: Isobutane/2-butene molar ratio of 15, 25 bar total pressure and a WSHF, referred to 2-butene, of 2 h^{-1}.

Table 9.7, it is difficult to obtain high alkylation conversions on commercial Nafion resins due to the very low surface area of the polymer. In order to increase this, one may decrease the 'particle size' of the Nafion polymer. This has been cleverly done by Hamer et al.[36] by partially depolymerizing the Nafion and dispersing this in a silica gel. By doing this, it has recently been reported that the activity in several reactions per unit weight of Nafion resin can be increased by using the high surface area Nafion resin/silica.[36] The high surface area Nafion resin/silica nanocomposites are also effective catalysts in isobutane/butene alkylation.[38] In Table 9.7 the catalytic results of Nafion polymer and that of the resin/silica nanocomposite (20wt% of Nafion, S_{BET} = 300 m^2 g^{-1}) are compared. It can be seen that not only the 2-butene conversion but also the selectivity to TMP is higher on the Nafion/silica nanocomposite catalyst. This higher selectivity to TMP is a consequence of a lower selectivity to C_{9+} hydrocarbons, while the nanocomposite catalyst shows a higher selectivity to cracked products (C_5–C_7) than pure Nafion. If we compare these results with those obtained on heteropolyacids (Table 9.5), it can be seen that Nafion shows a lower selectivity to TMP, a higher selectivity to C_9 and a lower selectivity to cracked products (C_5–C_7) than heteropolyacids.

9.4 Catalytic stability of solid acid catalysts

An important limitation for the use of solid acid materials as catalysts in the isobutane alkylation reactions is related to their stability during the reaction conditions. Figure 9.4 shows the variation of the yield to TMP with time on stream (TOS) obtained on different zeolites at 80 °C. It can be seen that although Mordenite appears to be the most active catalysts at low time on stream, Beta zeolite is the most effective taking into account the stability.

Figure 9.5 shows the variation of the conversion of 2-butene on strong solid acid catalysts at the best reaction conditions for each sample. It can be seen that although the strong solid acids are initially more active than Beta zeolite, they show a lower productivity at high TOS. In this way, the catalytic activity decrease with time on stream in the order: H-BETA > Nafion > $K_2H_{0.5}PW_{12}O_{40}$ > SZ.

Figure 9.6 shows the variation of the selectivity to TMP with the TOS at the same reactions conditions as those of Fig. 9.5. It can be seen that,

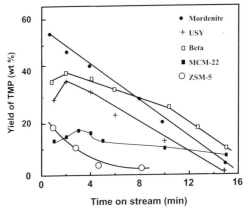

Fig. 9.4 Variation of the yield of TMP with the time on stream during the isobutane/2-butane alkylation on zeolite catalysts at 50 °C. Experimental conditions: Isobutane/2-butene molar ratio of 15, 25 bar total pressure and a WSHF, referred to 2-butene, of 2 h^{-1}.

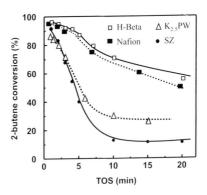

Fig. 9.5 Variation of the 2-butene conversion with the time on stream during the isobutane/2-butene alkylation stream on H-beta and superacid solid catalysts. Reaction temperatures: H-beta (50 °C); Nafion/SiO$_2$ (80 °C), K$_{2.5}$H$_{0.5}$PW$_{12}$O$_{40}$ (80 °C); and SZ (50 °C).

except in the case of SZ catalysts, the selectivity to TMP decreases with TOS. The different influence of the time on stream on the selectivity to TMP observed with SZ catalysts can be explained on the basis of the different acid strength of its active sites. In fact, the concentration of

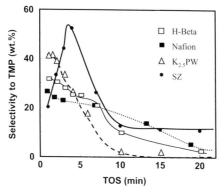

Fig. 9.6 Variation of the selectivity to TMP with the time on stream during the isobutane/2-butene alkylation on H-beta and superacid solid catalysts. Experimental conditions as Fig. 9.5.

TMPs sharply increases during the first minutes of TOS due to a strong decrease of the cracking activity as a consequence of the selective poisoning of the superacid sites. While this occurs, the sites with medium-strong acidity remaining are still able to carry out the alkylation reaction, and the selectivity to TMP is still high and a maximum selectivity to TMP is achieved at about 4 min. However, at higher times on stream, the selectivity to TMP decreases and this occurs at any of the reaction temperatures studied.

Solid acid catalysts generally suffer from a relatively rapid catalyst decay, requiring frequent catalyst regeneration. This requirement leads to a large and complex process system. Several integrated solutions have been reported. Besides that of the Topsoe's process,[50] Dejong et al.[51] have introduced a low amount of a metal on Beta zeolite to produce the hydrogenation and desorption of the heavier products adsorbed. The use or recycle, CSTR[52] or riser type reactors[53] have been reported to provide and alkylate to a quality comparable to that produced by conventional process.

It has also been reported that by working with CO_2 at supercritical conditions and at temperatures lower than the critical temperature of isobutane (< 135 °C) the catalyst life of microporous zeolitic (USY) and mesoporous solid acid (sulfated zirconia) catalyst could be prolonged.[54] Nevertheless, catalyst deactivation will still occur in a relatively short time, and this requires major efforts for the continuous regeneration of solid catalysts.

9.5 Conclusion

It has been reported here that besides supported liquid acids, strong solid acids can be used as isobutane/butene alkylation catalysts with good performance from the point of view of activity and selectivity. An important factor for the optimization of the catalytic behavior is the control of acidity and accessibility. This will change the relative selectivity to alkylation–cracking–oligomerization. However, the still unresolved problem is whether to prolong the life of the catalyst or to continuously regenerate the catalyst in an efficient way. Finally, one has to take into account that the price of the alkylate is an important parameter and limits the use of highly sophisticated catalysts and processes.

Acknowledgment

Financial support by the Comision Interministerial de Ciencia y Tecnología of Spain (Project MAT 97-0561) is gratefully acknowledged.

References

1. Corma, A.; Martinez, A. *Catal, Rev.-Sci. Eng.* 1993, **35**, 483–570.
2. Anastas, P.T.; Williamson, T.C., *Green Chemistry. Designing Chemistry for the environment*, ACS Symp. Series, vol. 625, ACS, Washington D.C. 1996.
3. a) OGJ Special Reports. *Oil and Gas Journal* 1996, Jan. 8, 29–32; b) OGJ Special Reports. *Oil and Gas Journal* 1994, Aug. 22, 49–51.
4. Rhodes, A.J. *Oil and Gas Journal* 1994, Aug. 22, 56–59.
5. Rhodes, A.J. *Oil and Gas Journal* 1994, Aug. 22, 52–55.
6. Sheckler, J.C.; Hammershaimb, H.U.; Ross, L.J.; Comey Ill, K.R. *Oil and Gas Journal* 1994, Aug. 22, 60–63.
7. Olah, G.A.; Batamack, P.; Deffieux, D.; Török, B.; Wang, Q.; Molnár, A.; Prakash, G.K.S. *Appl. Catal. A: General* 1996, **146**, 107–117.

8. Benazzi, E.; Joly, J.F.; Latieule S.; Marcilly, Ch. *Preprints of the American Chemical Society Symposia, Division of Petroleum Chemistry*, vol. **41**. American Chemical Society, Washington, 1996, 711–716.
9. Hommeltoft, S.I.; Ekelund, O.; Zavilla, J. *Ind. Eng. Chem. Res.* 1997, **36**, 3491–3497.
10. Cornet, D.; Goupil, J.-M.; Szabo, G.; Poirier, J.-L.; Clet G. *Appl. Catal. A: General* 1996, **141**, 193–205.
11. Chauvin, Y.; Hirschauer, A.; Olivier, H. *J. Molec. Catal.* 1994, **92**, 155–165.
12. Guo, C.; Liao, S.; Qian, Z.; Tanabe, K. *Appl. Catal. A: General* 1994, **107**, 239–248.
13. Corma, A. *Chem. Rev.* 1995, **95**, 559–614.
14. Venuto, P.B. *Microporous Mater.* 1994, **2**, 297–411.
15. a) Corma, A.; Martinez, A.; Martinez, C. *J. Catal.* 1994, **146**, 185–192. b) Corma, A.; Martinez, A.; Martinez, C. *Appl. Catal. A: General* 1996, **134**, 169–182.
16. Simpson, M.; Wei, J.; Sundaresan, S. *Green Chemistry: Designing Chemistry for the Environment*, Nastas, P.T. and Williamson, T.C. Eds., ACS Symp. Series, 1996, vol. **626**, 105–115.
17. Weitkamp, J. *Stud. Surf. Sci. Catal.* 1980, **5**, 65–75.
18. Stöcker, M.; Mostad, H.; Rorvik, T. *Catal. Lett.* 1994, **28**, 203–209.
19. Rorvik, T.; Mostad. H.B.; Karlsson, A.; Ellestad, O.H. *Appl. Catal. A: General* 1997, **156**, 267–283.
20. a) Corma, A.; Gómez, V.; Martinez, A. *Appl. Catal. A: General* 1994, **119**, 83–96. b) Corma, A.; Martinez, A.; Arroyo, P.A.; Monteiro, J.L.F.; Sousa-Aguiar, E.F. *Appl. Catal A: General* 1996, **142**, 139–150.
21. Corma, A.; Juan Rajadell, M.I.; López Nieto, J.M.; Martinez, A.; Martinez, C. *Appl. Catal. A* 1994, **111**, 175–189.
22. Child, J.E.; Huss, A.; Kenedy, C.R.; Miller, D.O.; Tabak, T.A. US Patent 1991, 5 012 033.
23. Collins, N.A.; Child, J.E.; Huss, A. *Preprints of the American Chemical Society Symposia, Division of Petroleum Chemistry* (vol. 41). American Chemical Society, Washington. 1996, 706–710.
24. Hino, M.; Arata, K. *J. Am. Chem. Soc.* 1979, **101**, 6439–6441.
25. Hino, M.; Arata, K. *J. Chem. Soc., Chem. Comm.*, 1980, 851–852.
26. Arata, K. *Adv. in Catal.* 1990, **37**, 165–211.
27. Song X.; Sayari, A. *Catal. Rev.-Sci. Eng.* 1996, **38**, 329–412.
28. Arata, K. *Appl. Catal.: A General* 1996, **146**, 3–32.
29. Okuhara, T.; Mizuno N.; Misono, M. *Adv. Catal.* 1996, **41**, 113-252.

30. Kozhevnikov, I.V. *Catal. Rev.-Sci. Eng.* 1995, **37**, 311–352.
31. Na, K.; Okuhara T.; Misono. M. *J. Chem. Soc., Chem. Comm.* 1993, 1422–1423.
32. Conolly, D.J.; Gresham, W.F. (1966) US Patent 3,282,875.
33. Sondheimer, S.J.; Bunce, N.J.; Fyfe, C.A. *JMS-Rev. Macromol. Chem. Phys.* 1986, **C26**, 353–413.
34. Olah, G.A.; Iyer, P.S.; Prakash, G.K.S. *Synthesis* 1986, 513–531.
35. Chakrabati A.; Sharma, M.M. *React. Polym.* 1990, **20**, 1–45.
36. a) Harmer, M.A. WO Patent, 1995, 95/19222. b) Harmer, M.A.; Farneth, W.E.; Sun, Q. *J. Am. Chem. Soc.* 1996, **118**, 7708–7715.
37. Palinko, I.; Rörök, B.; Prakash, G.S.K., Olah, G.A. *Appl. Catal. A: General* 1998, **174**, 147–153.
38. Botella, P.; Corma, A.; López Nieto, J.M. *J. Catal.* 1999, **185**, 371–377.
39. Guo, C.; Yao, S.; Cao, J.; Qian, Z. *Appl. Catal. A: General* 1994, **107**, 229–238.
40. Corma, A.; Juan Rajadell, M.I.; Löpez Nieto, J.M.; Martinez, A.; Martinez, C. Abst. 106th *ACS Meeting, Division of Colloid & Surface Chemistry* American Chemical Society, Chicago (1993), paper-coll. 208.
41. Corma, A.; Fornés, V.; Juan Rajadell, M.I.; López Nieto, J.M. *Appl. Catal. A: General* 1994, **116**, 151–163.
42. Corma, A.; Martinez, A.; Martinez, C. *J. Catal.* 1994, **149**, 52–60.
43. Hess, A.; Kemnitz, E. *Appl. Catal. A: General* 1997, **149**, 373–389.
44. Rajadell, M.I.J. Ph.D. Thesis, Universidad de Valencia, Burjassot, 1995.
45. Corma, A.; Martinez, A.; Martinez, C. *Appl. Catal. A: General* 1996, **144**, 249–268.
46. Corma, A.; Martinez, A.; Martinez, C. *J. Catal.* 1996, **164**, 422–432.
47. Blasco, T.; Corma, A.; Martinez, A; Martinez-Escolano, P. *J. Catal.* 1998, **177**, 306–313.
48. Essayem, K.; Kieger, S.; Coudurier, G.; Vedrine, J.C. *Stud. Surf. Sci. Catal.* 1996, **101**, 591–600.
49. Rorvik, T.; Dahl, I.M.; Mostad, H.B.; Ellestad, O.H. *Catal. Lett.* 1995, **33**, 127–134.
50. Hommeltoft, S.I. *Prepr. of the ACS Symp., Division of Petroleum Chemistry* (vol. 41). American Chemical Society, Washington. 1996, 700–705.
51. Dejong, K.P.; Mesters, C.M.A.M.; Peferoen, D.G.R.; Vanbrugge, P.T.M.; Degroot, C. *Chem. Eng. Sci.* 1996, **51**, 2053–2060.

52. Rao P.; Vatcha, SR. *Prepr. of the ACS Symp., Division of Petroleum Chemistry* (vol. 41). American Chemical Society, Washington. 1996, 685–691.
53. Barger, P.T., Frey, S.J., Gosling, C.D., Sheckler, J.C., Prepr. of the ACS Symp., Division of Petroleum Chemistry (Vol. 44). American Chemical Society, Washington, 1999, 134–140.
54. Clark, M.C.; Subramaniam, B. *Ind. Eng. Chem. Res.* 1998, **37**, 1243–1250.

10 Selective catalytic oxidation of 1,2-diols in alkaline solution: an environmentally friendly alternative for α-hydroxy-acids production

Laura Prati and Michele Rossi

Abstract

This article reviews our recent studies on the selective, liquid phase catalytic oxidation of 1,2-diols to α-hydroxy carboxylates involving dioxygen as the oxidant; well known VIII group metal catalysts have been compared to innovative gold based catalysts. Depending slightly on the catalyst Pd/C, Pt/C and Au/C show, in alkaline solution under mild conditions (T = 343–363 K, pO_2 = 300 kPa), very high selectivity (90–100%) toward the mono-oxygenation of ethane-1,2-diol and propane-1,2-diol, even at high diol conversion (80–94%). However with respect to the classical Pt and Pd based catalysts, gold on carbon is quite stable against aging as proven by recycling tests. Thus, the performances of Au/C suggest its use as the catalyst for selective oxidation in the synthesis of α-hydroxy acids representing an environmentally friendly oxidative methodology.

10.1 Introduction

In organic chemistry, oxidation is a very important reaction in the manufacture of both bulk and fine chemicals. However, increasing environmental awareness has made classical stoichiometric oxidation, that produces large amounts of waste, undesirable. Moreover, given the importance of not only the amount but also of the nature of wastes, as recently outlined,[1] a great interest has been shown in catalytic oxidation using dioxygen as a clean oxidant, with water as its coproduct.

Two different methodologies for catalytic oxidation involving dioxygen are possible: the first, the most attractive from an industrial point of view, is represented by a continuous gas phase reaction; the second involves the use of a solvent and it is used when the gas-phase methodology is unsuitable for high boiling points and/or low thermal stability of the reagents or products. However, short residence times on fixed-bed catalysts can sometimes overcome thermal instability in polyfunctionalized organic molecule. Thus, for example, control of the catalyst–reactant contact time allows isoprenol, in the presence of a silver on silica catalyst (a key step of the BASF synthesis of citral), to be oxidized to isoprenal at 500 °C, and ethylene glycol, in the presence of silver on SiC, to be dehydrogenated to glyoxal at 450 °C.[2]

In liquid phase reactions the conditions are normally milder. In fact, a typical liquid phase application is the oxidation over noble metal catalysts of glucose to gluconic acid,[3] widely studied as a possible alternative to the currently used enzymatic process.[4] However, despite the great effort put into upgrading the activity and selectivity of supported platinum group metal catalysts, these liquid phase oxidations have not, until recently, been industrially applied, since the enzymatic route is still more convenient.

The principal reason is a significant catalyst deactivation, the mechanism of which is not yet fully understood. It has been reported that the presence of strongly adsorbed reaction products or by-products reduces catalytic active sites and, consequently, the activity of the catalyst, but also recognized is a poisoning effect of the oxygen itself, through the formation of less active oxidized species.[5] It is commonly accepted that the mechanism of metal catalyzed oxidations of alcohols is a dehydrogenative one[6] where both dioxygen and substrate are adsorbed on the catalyst. It is the balance between the dehydrogenation of the substrate and the oxidation of the hydride by adsorbed oxygen that determines the efficiency of the catalyst. If the oxygen consumption rate is slower than the adsorption of O_2 on the catalyst surface, a progressive deep oxidation of the metal takes place and the catalyst deactivates.

Experimental solutions have been proposed, at least to reduce the oxygen poisoning, for example the use of low partial pressure of oxygen, low stirring speeds and 'diffusion-stabilized' carriers,[7] but there has been a marked enhancement of catalyst resistance, especially in the field of carbohydrate oxidations, by introducing a heavy metal such as Bi or Pb as a modifier of Pt or Pd on carbon catalysts.[8] Several studies on this subject have not yet fully clarified the real role of the second metal in enhancing both selectivity and catalyst life. On the contrary, less investigated are the new catalytic materials showing intrinsic high selectivity and catalytic resistance to chemical poisoning. This article reports our studies on this different approach dealing with a new class of gold-based catalysts and their application to the catalytic oxidation of vicinal diols in water solution.[9] In particular, we discuss here the selective oxidation of ethane-1,2-diol and propane-1,2-diol to glycolic acid (C2) and lactic acid (C3). Such oxidation processes could represent a valuable target, providing a clean methodology that uses oxygen in water solution for the synthesis of industrially important α-hydroxy-acids.

10.2 Experimental

10.2.1 Reagents and apparatus

Reagents were of the highest purity from Fluka and were used without any further purification. R(-)-propylene glycol was from Fluka showing $[\alpha]_D$ $-16.5 \pm 1°$. NaOH and D_2O were 99.9% pure from Merck and stored under nitrogen. NaOD was 40% in D_2O (> 99.5 atom % D) from Fluka. Gaseous oxygen from SIAD was 99.99% pure. Gold powder was of the highest purity grade from Fluka. Commercial 5% Pd/C was supplied by Süd Chemie, 5% Pt/C by Engelhard.

Activated carbon (5–100 μ) had a specific area of 1200 m^2 g^{-1}; Al_2O_3 from La Roche had a specific area of 420 m^2 g^{-1}, SiO_2 (100–150 μ) had a specific area of 419 m^2 g^{-1} and TiO_2 (P25, 80% anatase) was from Degussa.

Reactions were carried out at the appropriate temperature in a thermostatted glass reactor (30 ml) provided with an electronically controlled magnetic stirrer connected to a large reservoir (5000 ml) containing oxygen at 300 kPa. The oxygen uptake was followed by a mass flow controller connected to a PC through an A/D board, plotting a flow/time diagram.

10.2.2 Catalyst preparation

Gold catalysts were prepared using a $HAuCl_4$ 0.1M solution obtained by dissolving 1.97 g of gold powder in a minimum amount of a 3 : 1 (v/v) mixture of concentrated HCl and HNO_3 and then diluted to 100 ml with distilled water.

1% Gold on silica was prepared by using incipient wetness impregnation and reducing the catalyst by calcinating the catalyst at 523 K for 4 h in air.

1% Au on α-Fe_2O_3 was prepared by the coprecipitation method as previously reported.[10]

1% Au on Al_2O_3 was prepared by using either incipient wetness impregnation or deposition–precipitation method previously reported.[11] The reduction step was performed by calcinating the catalyst at 523 K for 4 h in air.

1% Gold on carbon catalysts were prepared by two different methodologies:

1. Incipient wetness impregnation: the support (2 g) was impregnated with 1 ml of 0.1M $HAuCl_4$ diluted with distilled water to a volume equal to the pore volume of the support. The suspension was mixed for 20 min then added to a hot solution of HCOONa (20 ml of water and 200 mg of sodium formate).
2. Deposition–precipitation method: the solution of $HAuCl_4$ 0.1M (1 ml) was diluted with distilled water (10 ml) and a saturated solution of Na_2CO_3 was added until a fixed pH of 10 was reached. Then the mixture was added to a stirred slurry of carbon (2 g) in distilled water (20 ml). The slurry was allowed to stand for 1 h then HCHO 37% (1.5 ml) was added.

After reduction, all the catalysts were filtered and then washed with hot water until the filtrate was chloride free. The catalysts were used in wet form.

10.2.3 XRPD

X-ray diffraction experiments were performed on a Rigaku D III-MAX horizontal-scan powder diffractometer with Cu-Kα radiation, equipped with a graphite monochromator in the diffracted beam. Crystallite sizes of gold were estimated from peak half-widths by using Scherrer's equation with corrections for instrumental line broadening.

10.2.4 Oxidation procedures

10.2.4.1 Oxidation under alkaline conditions

The substrate (0.50 g, 8.06 mmol), NaOH (0.350 g, 8.75 mmol) and eventually the catalyst (diol/metal = 1000) were mixed in distilled water (total volume 10 ml). The reactor was pressurized at 300 KPa of O_2 or alternatively of N_2 and thermostatted at the appropriate temperature. The mixture was stirred and the samples analyzed at various times by HPLC and ^{13}C-NMR.

10.2.4.2 Oxidation of propane-1,2-diol at pH 8

The substrate (0.52 g, 6.8 mmol) and the catalyst (diol/M = 1000) were mixed in distilled water (total volume 10 ml). Dioxygen was bubbled at atmospheric pressure into the stirred mixture thermostatted at the appropriate temperature. A solution of NaOH 0.1M was added dropwise to maintain the pH of the solution at 8. The pH was measured by a Metrohom 744 pHmeter. Samples were analyzed at various times by HPLC.

10.2.4.3 Recycling tests

After the first run the catalyst was filtered off and reused, without any further purification, in the next run with a freshly prepared solution of glycol. After four runs the catalyst was filtered and washed several times with water until neutral pH of water. The metal content was determined by ICP analysis on a Jobin Yvon JY24.

10.2.5 H-D exchange experiments

All the reagents were handled under N_2 after exchange of labile hydrogen (OH) which was performed by dissolving the substrates in D_2O and then evaporating the solution. The reagent was then redissolved in D_2O and allowed to react with O_2 following the same oxidation procedure as reported above except that NaOD was used instead of NaOH. ^{1}H and ^{13}C-NMR spectra were registered and the solution evaporated. The D-NMR spectrum was registered by dissolving the residue in H_2O and the D-H exchange evaluated by comparison of ^{1}H and D-NMR spectra.

10.2.6 Analysis of products

The products were identified by comparison with authentic samples. Quantitative analyses were performed by either HPLC (Alltech OA-1000 column (300 mm × 6.5 mm); aqueous H_2SO_4 0.01 M (pH 2.1) (0.8 ml min^{-1}) as the eluent) or ^{13}C-NMR (300 MHz) methods, using an internal standard (propionic acid).

10.3 Results and discussion

10.3.1 Catalysts

As reported in the Introduction, conventional catalysts used in the liquid phase oxidation of alcohols are platinum or palladium on carbon. According to Mallat and Baiker,[5] the introduction of promoters such as Bi or Pb has led to highly selective oxidations. In particular, the oxidation of glucose to gluconic acid (C1 oxidation) can be performed with high chemo-selectivity at high conversion in the presence of a Pt/Bi on carbon catalyst, whereas oxidizing simple polyols in the presence of the same catalyst results in a high selectivity toward the secondary alcoholic function with respect to the primary one.[12] Thus, a scale of reactivity in the presence of oxygen and modified bi- or trimetallic catalysts seems to be CHO > CHOH > CH_2OH. Along with high selectivity, there was also an increased catalyst life but the reason for this beneficial effect of promoters on platinum or palladium catalysts is not yet fully understood.

Among the noble metals, platinum seems to be the least easily poisoned by over-oxidation, followed by Ir, Pd, Rh and Ru, the last being almost inactive.[13] This trend agrees with the minor oxidability of metals correlated to higher redox potential.[14]

Some surface studies dealing with the adsorption of oxygen on supported silver/gold alloy have revealed that gold inhibits subsurface oxygen diffusion and decreases the presence of atomic oxygen compared to the molecular one.[15] This particular behavior, in addition to a high redox potential, suggested investigating gold catalysts as possible candidates for liquid phase oxidation, less prone to deactivation by over-oxidation than platinum or palladium catalysts.

On the other hand, gold has recently been discovered as active metal for gas phase oxidation. It is worth noting that specific preparation methods allowing a high dispersion of the metal on the support has been the basis of these applications.[16] Furthermore, it has been reported that, for gold catalysts, activity and selectivity strongly depend on the preparation method, type of support and metal size particle distribution. Most of the gas phase applications of gold catalysts concern CO oxidation and hydrocarbon combustion, whereas no application in liquid phase processes (apart from some examples in electrocatalysis) has been reported. Therefore, in looking for possible uses of gold in liquid phase oxidation, our investigations started from the preparation of various gold-based catalysts and catalytic tests on simple diols under mild conditions.

The deposition–precipitation method has been indicated as the best way of preparing dispersed gold nanoparticles;[11,16] also for α-Fe$_2$O$_3$ the coprecipitation route gives good results.[10] It has also been reported that amorphous supports produce larger gold particles than crystalline supports and that the activity of gold is greatly enhanced with particles having less than 10 nm diameter. Therefore, we could expect some difficulties in supporting gold on activated carbon, a typical amorphous support. An XRPD study conducted on gold deposited on carbon according to two different preparation methods (incipient wetness impregnation and deposition-precipitation) revealed smaller gold particles using the latter method (Table 10.1). Even if the XRPD technique determines the crystallite size (coherently scattering domains) that does not necessarily correspond to the catalytically active particles, these data can be used for comparative purposes. As a test reaction we used the oxidation of ethane-1,2-diol, under standard conditions (343 K, 300 kPa O$_2$) in alkaline solution, and found that both the activity (TOF)

Table 10.1 Influence of preparation methods of 1%Au on carbon.

Preparation method	XRPD mean gold crystallite diameter (nm)	TOF mol conv. mol Au^{-1} h^{-1}	select. % to GLA at 80% conv.
Incipient wetness impregnation	12	325	86
Deposition–precipitation	7	405	93

Reaction conditions: ethane-1,2-diol/Au = 1000; NaOH/ethane-1,2-diol = 1; T = 70 °C; pO$_2$ = 300 kPa. GLA = Glycolate. TOF numbers were calculated on the basis of total gold.

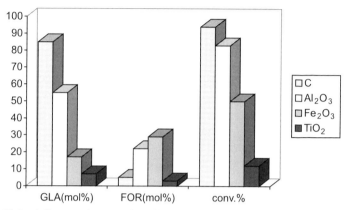

Fig. 10.1 Influence of support in gold catalyzed oxidation of ethane-1,2-diol. Reaction conditions: ethane-1,2-diol/M = 1000; NaOH 2.6M; NaOH/ethane-1, 2-diol = 1; T = 70 °C; pO_2 = 300 kPa. GLA = glycolate; FOR = formate. Na_2CO_3 was also formed. Oxalate was not detected.

and selectivity toward glycolate were superior in the case of gold prepared by the deposition–precipitation method (Table 10.1). These results were then compared to those obtained, under the same conditions, using different supports, namely Al_2O_3, TiO_2 and α-Fe_2O_3, which produce catalysts active in gas-phase oxidation.

As reported in Fig. 10.1, oxidic supports show a minor activity and selectivity compared to carbon. Interestingly, in the case of the alumina catalyst, even the XRPD measurements revealed gold crystallites comparable with respect to carbon, gold on alumina shows a minor activity and selectivity. TiO_2 and α-Fe_2O_3 produced practically inactive catalysts. Therefore, we can highlight a peculiar effect of carbon on the behavior of gold catalysts in the oxidation of ethane-1,2-diol in alkaline solution and establish the benefical effect of using the deposition–precipitation method in obtaining small and active gold particles.

10.3.2 Ethane-1,2-diol oxidation

Currently used chemical methods for producing glycolic acid (hydroxyacetic acid) involve either the reaction of formaldehyde with carbon monoxide and water in the presence of an acidic catalyst as HF (Du Pont Chemicals and Hoechst) (1), or the hydrolysis of chloroacetic acid (2):

1) $HCHO + CO + H_2O \rightarrow HOCH_2COOH$ at high pressure and temperature

2) $ClCH_2COO^- + OH^- \rightarrow HOCH_2COO^-$ at 90–130 °C

Although the conditions used in (1) are not disclosed, the pressure is believed to range between 100 and 500 atm and the temperature between 50 and 100 °C. The alternative hydrolysis of molten chloroacetic acid with 50% sodium hydroxide is carried out at 90–130 °C (2).[17] In accordance with the current tendency to clean chemical processes by using clean methodologies (green chemistry),[1] we studied an alternative method, of general validity, for the synthesis of α-hydroxyacids, for the synthesis of glycolic acid based on selective catalytic oxidation of ethylene glycol with oxygen in water solution.

In the class of vicinal diol, the first member ethane-1,2-diol presents some peculiarities: the presence of a double primary alcoholic function and the facile occurrence of C–C bond cleavage under oxidative conditions. In fact, the application of platinum and palladium on carbon catalysts is limited by over-oxidation, and the patent literature reports mostly catalytic processes on protected substrates such as methoxy glycol and polyethylene glycol.[18] As already mentioned, doping with heavy metal palladium or platinum on carbon allows better performance (95% selectivity to glycolic acid).[19] However, the most notable result in glycolic acid production from ethane-1,2-diol was, to the best of our knowledge, claimed to be from the use of a different metal, namely Ir/C operating at 10 atm and 80 °C (87% selectivity at 98% conversion).[20]

On the contrary, different catalysts (CuO/ZnO or Ag/SiC) in gas phase processes normally produce the oxidation of both the functionality of ethane-1,2-diol, a valuable method to produce glyoxal.[2] The results obtained using new gold catalysts in the mono-oxygenation of ethane-1,2-diol under mild conditions are seen to be attractive by comparing gold catalyst performance with more conventional catalysts such as Pd/C and Pt/C, and others such as Ir/C[20] and Cu/C[21] reported in the literature (Fig. 10.2).

Although commercial 5% platinum on carbon and 5% palladium on carbon showed higher activity than 1% gold on carbon, they were affected by a large over-oxidation to C–1 products (formate and carbonate). Nevertheless, by operating at 50 °C a selectivity of about

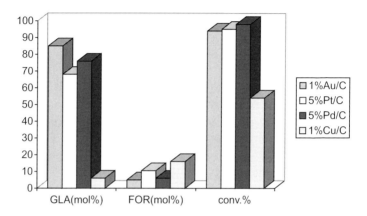

Fig. 10.2 Oxidation of ethane-1,2-diol with various catalysts. Reaction conditions: ethane-1,2-diol/M = 1000; NaOH 2.6M; NaOH/ethane-1,2-diol = 1; pO_2 = 300 kPa. GLA = glycolate; OXA = oxalate; FOR = formate. Carbonate was also formed. Reaction temperatures: 70 °C for gold and copper catalysts; 50 °C for palladium and platinum catalysts. Reaction times: 3 h for gold catalyst; 2 h for platinum and palladium catalysts; 6 h for copper catalyst.

70% was obtained (Fig. 10.2).[9] Copper on carbon produced relevant quantities of formate even at low conversion owing to the known activation toward C–C bond cleavage[21] and Ir on carbon catalyst, prepared according to the literature,[20] was inactive under the tested reaction conditions (Fig. 10.2).

It has been reported that catalyst life also depends on the basicity of the reaction media,[5] deactivation being reduced by operating at high pH. In recycling tests, where the catalysts were reused several times, we found that 5% Pd/C shows a marked leaching of metal (45% after four cycles), whereas in the case of 5% Pt/C there was, besides a trascurable transfer of platinum, a slight drop in selectivity (–31%). On the contrary, 1% Au/C revealed an almost constant activity and selectivity during seven cycles.

A second benefit of operating in alkaline solution is that the resulting selectivity can be improved. In fact, some equilibria can co-operate to produce glycolate from different precursors. Starting from ethane-1,2-diol let us consider a reaction pattern where glycolate can be formed either by oxidation of the intermediate glycolic aldehyde (b) or by the

Canizzaro reaction of the intermediates glyoxal (intramolecular) (c) and glyoxylate (intermolecular) (f):

a) $HOCH_2CH_2OH$ $+[O]$ $\rightarrow$ $HOCH_2CHO$
b) $HOCH_2CHO$ $+[O]$ $\rightarrow$ $CHOCHO, HOCH_2COO^-$
c) $CHOCHO$ $+OH^-$ $\rightarrow$ $HOCH_2COO^-$
d) $HOCH_2COO^-$ $+[O]$ $\rightarrow$ $CHOCOO^-$
e) $CHOCOO^-$ $+[O]$ $\rightarrow$ $(COO^-)_2 \rightarrow CO_3^{2-}, HCOO^-$
f) $2CHOCOO^-$ OH^- $\rightarrow$ $(COO^-)_2 + HOCH_2COO^-$

$$HOCH_2CH_2OH + [O] + OH^- \rightarrow HOCH_2COO^- + (COO^-)_2 + CO_3^{2-} + HCOO^-$$

In the case of the gold catalyzed oxidation, the contribution of the intermolecular Canizzaro reaction to the formation of glycolate is thought to be negligible as the absence of oxalate in the oxidation of either ethane-1,2-diol or glycolate excludes the presence of glyoxylate as an intermediate.

The high selectivity shown by gold catalysts could also be correlated to the particular resistance of glycolate to further oxidation, or to the poor tendency of gold to catalyze the carbon–carbon bond scission. In fact, a solution of sodium glycolate is quite stable under oxidative conditions in the presence of the gold catalyst whereas it is oxidized in the presence of both palladium and platinum catalysts to produce relevant amounts of carbonate and oxalate (Fig. 10.3). Therefore the carbonate or formate can, in the case of a Pd or Pt catalyzed oxidation of ethane-1,2-diol, derive form the C–C bond scission of oxalate, but in the case of gold their formation occurs directly from glycol as glycolate is stable. In fact, with gold, the selectivity of the reaction is almost constant at any conversion.

In conclusion, the selectivity shown by Pd, Pt and Au in the catalyzed oxidation of ethane-1,2-diol in alkaline solution is the result of a synergic effect of OH^- in catalyzing different reactions that affect the intrinsic selectivity of the catalyst. Using ethane-1,2-diol as the reagent, the separation of these two components is extremely difficult as the base-catalyzed reactions on glyoxal and glyoxylate also take place at relatively low pH (7.5–8), and below these values the rate of O_2 adsorption becomes quite low. Thus, we studied the effect of basic conditions on the oxidation of another substrate having intermediates more resistant to basic conditions.

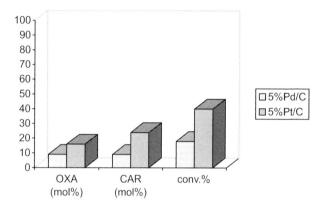

Fig. 10.3 Oxdiation of glycolate. Reaction conditions: glycolic acid/M = 1000; NaOH/glycolic acid = 1; pO$_2$ = 300 kPa; T = 70 °C; t = 2h; OXA = oxalate; CAR = carbonate.

10.3.3 Propane-1,2-diol oxidation

The current chemical route for lactic acid production uses the lactonitrile route (i).[17] It involves the base-catalyzed addition of hydrogen cyanide to acetaldehyde to produce lactonitrile which is purified by distillation and then hydrolyzed by using hydrochloric or sulfuric acid. Other syntheses of lactic acid (ii, iii) have been proposed but they have not found industrial application.

i) $CH_3CHO + HCN \rightarrow CH_3CHOHCN$
  $CH_3CHOHCN + 2H_2O + \frac{1}{2}H_2SO_4 \rightarrow CH_3CHOHCOOH + \frac{1}{2}(NH_4)_2SO_4$
ii) $CH_3CHO + CO + H_2O \rightarrow CH_3CHOHCOOH$ at elevated T and P
iii) $CH_3ClCHCOO^- + OH^- \rightarrow CH_3CHOHCOO^-$

Apart from problems connected with the handling of a dangerous reagent, such as HCN, a limitation encountered in this lactic acid production is its purification, which is accomplished by transforming the crude acid in the corresponding methyl ester followed by distillation and acid-catalyzed hydrolysis of the purified lactic acid. Similar problems arise in the classical fermentation process (iv) which also has the shortcoming of low productivity.

iv) $C_6H_{12}O_6 + Ca(OH)_2 \rightarrow (CH_3CH_2HOCOO)_2Ca + 2H_2O$
$(CH_3CH_2HOCOO)_2Ca + H_2SO_4 \rightarrow 2CH_3CH_2HOCOOH + CaSO_4$

Value-added applications (for example bio-polymers) require purified lactic acid, obtainable, as in the case of the synthetic one, by esterification, distillation and subsequent hydrolysis. In addition, the lactic acid preferred for many applications has to be optically active, something that is difficult to obtain via the fermentation route.

Despite the many studies focused on the purification step (ultrafiltration, electrodialysis), little effort has been devoted to the search for new *green* processes for lactic acid production that could eventually be adapted to the production of an optically active acid. Accordingly to our earlier results, the selective catalytic oxidation of propane-1,2-diol could be a convenient alternative route. In this case there is (differing from that of ethane-1,2-diol) a primary and a secondary alcoholic function that introduce the problem of chemoselectivity, hydroxyacetone and lactic acid being possible products and both of commercial interest. Supported platinum group metals doped with bismuth have been reported to catalyze the oxidation of the secondary alcoholic function yielding hydroxyacetone,[12] whereas lead modified palladium on carbon catalyst has been claimed to be able to oxidize both the primary and the secondary alcohol, producing pyruvic acid.[22] On the contrary, palladium on carbon (a monometallic catalyst) has been reported to oxidize propane-1,2-diol in a non-selective manner.[22] Note that the same catalyst used in the oxidation of propane-1,3-diol (a double primary alcohol less prone to C–C bond scission than ethane-1,2-diol) produces 3-hydroxy-propanoic acid in high yield (86%).[23]

In alkaline solution at 70 °C under 300 kPa of O_2, we found that commercial 5% Pd/C and 5% Pt/C produce lactate in good selectivity (89–90%) at high glycol conversion (73–80%).[9] The gold catalyst, as in the case of ethane-1,2-diol, shows lower activity than palladium or platinum, but by operating at 90 °C instead of 70 °C, we were able to reach a comparable activity. In terms of selectivity, even by operating at higher temperature, we obtained a total selectivity toward lactate (at 78% conversion) (Table 10.2).

As outlined in the case of ethane-1,2-diol, the basic conditions could influence the final selectivity as possible intermediates in the reaction undergo base-catalyzed Canizzaro rearrangement. Thus, a careful

Table 10.2 Oxidation of propane-1,2-diol with various catalysts.

Catalyst	T (°C)	TOF mol conv. mol M^{-1} h^{-1}	Select.% to LA at 80% conv.
1%Au/C	90	780	100
5%Pt/C	70	730	89
5%Pd/C	70	800	90

Reaction conditions: NaOH/propane-1,2 diol = 1; diol/M = 1000; pO$_2$ = 300 kPa. LA = lactate.
M = metal. TOF numbers were calculated on the basis of total metal.

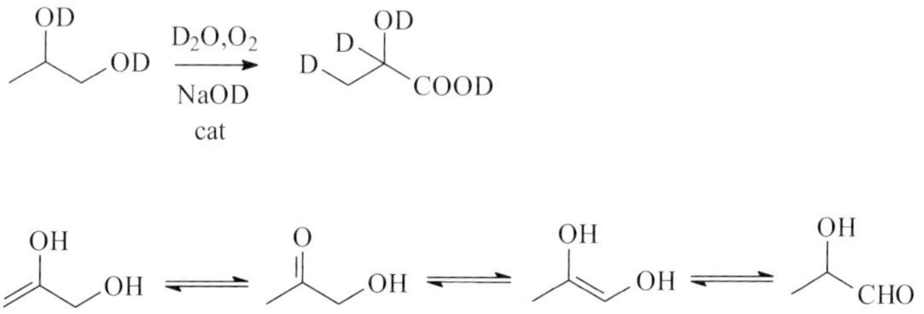

Fig. 10.4 H,D-exchange during propane-1,2-diol oxidation in deuterated reagents. Reaction conditions: NaOD/propane-1,2-diol = 1; diol M^{-1} = 1000; pO$_2$ = 300 kPa; T = 90 °C.

investigation was performed to clarify the role of the catalyst and that of OH$^-$ ions.

The oxidation of propane-1,2-diol conducted in the presence of deuterated reagents (NaOD in D$_2$O) produces lactate deuterated not only at the expected α-position but also at the methyl group (Fig. 10.4). In addition, a series of experiments, performed in deuterium oxide and sodium deuterium oxide, highlighted that neither sodium lactate nor the starting diol produce H–D exchange. A possible explanation for H–D exchange in the lactate at the C2 and the C3 positions is based on a mechanism involving the formation of hydroxyacetone as an intermediate, that in alkaline solution forms two different enols (Fig. 10.4). In fact, by oxidizing hydroxyacetone at 70 °C in alkaline solution with O$_2$ we obtained lactate in high yield. Only in the case of gold was the reaction totally selective whereas by using Pt/C or Pd/C a small amount of pyruvate was detected (2–3%). A probable reaction pathway for the oxidation of hydroxyacetone to lactate involves the formation of pyruvic

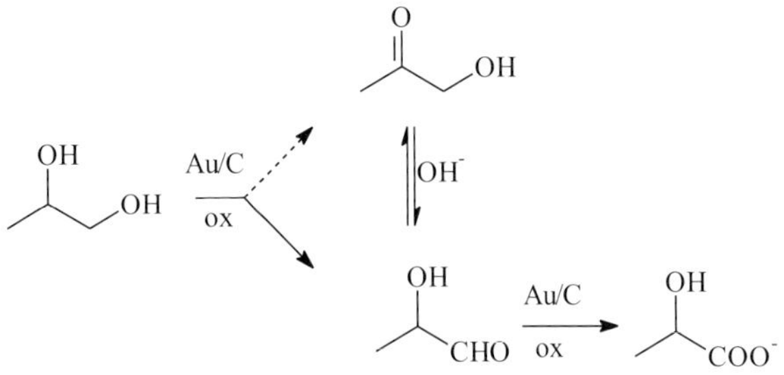

Fig. 10.5 Gold catalyzed oxidation of propane-1,2-diol under alkaline conditions.

aldehyde that rapidly undergoes an intramolecular Canizzaro reaction. In fact, pyruvic aldehyde, under basic conditions and in the absence of metal catalysts, quantitatively yields lactate, whereas under oxidative conditions, in the presence of Pd, Pt and Au catalysts, it gives, along with the lactate, variable amounts of pyruvate (2–4%). Therefore, the absence of pyruvate among the products in the gold-catalyzed oxidation of propane-1,2-diol excludes pyruvic aldehyde formation during the reaction and, consequently, lactate formation via the Canizzaro reaction. In the case of the gold-catalyzed oxidation, the presence of hydroxyacetone was demonstrated by H–D exchange experiments and it can be deduced that it is more rapidly oxidized to lactate (through lactaldehyde) than to pyruvic aldehyde (Fig. 10.5).

Moreover, in a more general reaction scheme for the catalyzed oxidation of propane-1,2-diol under basic conditions, we have to consider that lactate production can occur either by oxidation of the primary alcoholic function via lactaldehyde or by the intramolecular Canizzaro reaction of pyruvic aldehyde. This latter can be formed by oxidation of propandiol at the secondary hydroxyl group giving hydroxyacetone, but the lactaldehyde and hydroxyacetone could also be in equilibrium via their enols. Pyruvate, produced in the palladium or platinum-catalyzed oxidation, derives from hydroxyacetone as lactate is stable under oxidative conditions (Fig. 10.6). Unfortunately, the product distribution obtained in basic solution does not give any information on catalyst chemoselectivity, owing to the presence of equilibria that allow some intermediates to interconvert.

Thus, only by suppressing the base-catalyzed equilibria can we derive information on the real selectivity of the catalysts. Although the reaction

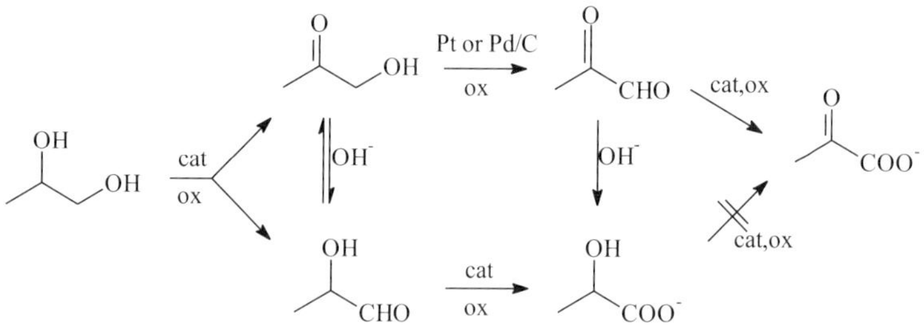

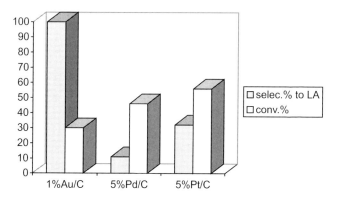

Fig. 10.6 General pathway of catalyzed oxidation of propane-1,2-diol.

Fig. 10.7 Oxidation of propane-1,2-diol at pH 8. Reaction conditions: diol/M = 1000; pO_2 = 100 kPa. The pH was maintained at 8 by dropping NaOH 0.1M. LA = lactate. T = 70 °C.

rates decrease with the pH, we were able, by operating at controlled pH (pH = 8) at which pyruvic aldehyde is stable and hydroxyacetone does not produce H–D exchange, to determine that gold on carbon was totally selective towards lactic acid, whereas platinum and palladium on carbon also produced relevant quantities of hydroxyacetone and pyruvate (Fig. 10.7). From these data, it can be deduced that gold shows an intrinsic high selectivity toward the oxidation of the primary hydroxyl, whereas Pd and Pt do not discriminate between primary and secondary hydroxyls.

Therefore, the high selectivities shown by palladium and platinum catalysts under strong basic conditions are due to OH^- catalyzed equilibria, favoring lactate formation that counterbalances the intrinsic low selectivity of these catalysts.

10.4 Conclusion

Our results indicate a new application for gold as an active metal for catalytic oxidation in the liquid phase. In particular, the high selectivity to mono-oxygenation achieved in the case of ethane and propane-1,2-diols suggests a new, suitable method for glycolic and lactic acid production of low environmental impact based on the use of dioxygen as the oxidant in water solution. For the first time there has been the clarification of the role of the base in enhancing the selectivity of catalytic systems based on palladium and platinum metals which do not show an intrinsic high selectivity. However, considering that metal leaching precludes the use of palladium and that platinum, on recycling, suffers over-oxidation leading to carbon–carbon bond scission, it can be concluded that gold on carbon is the best candidate for catalytic applications in the liquid phase oxidation of diols. Further studies are needed to develop a synthetic strategy to obtain optically active lactic acid.

References

1. Sheldon, R.A.; Dakka, J. *Catalysis Today*, **1994**, *19*, 215.
2. Gallezot, P.; Tretjak, S.; Christidis, Y.; Mattioda, G.; Schouteeten, A. *Journal of Catalysis*, **1993**, *142*, 729.
3. a) Fiege, H.; Wedemeyer, K. *Angewangte Chemie*, **1981**, *93*, 812. b) Fuertes, P.; Fleche, G.; Roquette Freres *E.P. 233816*, **1986**. c) Wenkin, M.; Touillaux, R.' Ruiz, P.; Delmon, B.; Devillers, M. *Applied Catalysis A: General* **1996**, *148*, 181 and references cited therein.
4. Hustede, H.; Haberstroh, H.J.; Schinzing, E. in *Ulmann's Encyclopedia of Industrial Chemistry* (B. Elvers, S. Hawkins, M. Ravenseroft, J.F. Rounsaville, G. Shulz, eds.) **1989**, vol. A12, p. 449, VCH, Veinheim.
5. Mallat, T.; Baiker, A. *Catalysis Today*. **1994**, *19*, 247 and references cited therein.
6. Di Cosimo, R.; Whitesides, G.M.; *Journal of Physical Chemistry*, **1989**, *93*, 768.

7. Van Dam, H.E.; Kieboom, A.P.G.; van Bekkum, H. *Applied Catalysis*, **1987**, *33*, 361.
8. a) Saito, H.; Ohnaka, S.; Fukuda, S. *European Patent* 0142725, **1984**. b) Despeyroux, B.M.; Deller, K.; Peldszus, E. *Studies on Surface Science and Catalysis*, **1990**, *55*, 159.
9. a) Prati, L.; Rossi, M. *Studies on Surface Science and Catalysis*, **1997**, *110*, 509. b) Prati, L.; Rossi, M. *Journal of Catalysis*, **1998**, *176*, 552.
10. Haruta, M.; Yamada, N.; Kobayashi, T.; Iijima,S. *Journal of Catalysis*, **1989**, *115*, 301.
11. Tsubota, S.; Cunningham, D.A.H.; Bando, Y.; Haruta, M. in *Preparation of Catalysts VI* (G. Poncelet, J. Martens, B. Delmon, P.A. Jacobs and P. Grange eds.) **1995**, p. 227, Elsevier, Amsterdam.
12. a) Garcia, R.; Besson, M.; Gallezot, P. *Applied Catalysis A: General*, **1995**, 127, 165. b) Kimura, H.; Tsuto, K. U.S. Patent 5,274,187, **1993**.
13. Van Dam, H.E.; Wisse, L.J.; Van Bekkum, H. *Applied Catalysis*, **1990**, *61*, 187.
14. Gallezot, P. *Catalysis Today*, **1997**, *37*, 405.
15. Kondarides, D.I.; Verykios, X.E. *Journal of Catalysis* **1996**, *158*, 363.
16. a) Haruta, M. *Catalysis Today*, **1997**, 36, 153. b) Haruta, M. *Catalysis Surveys of Japan*, **1997**, *1*, 61.
17. Datta, R. in *Encyclopedia of Chemical Technology*, Vol. 13, 4th edn, **1995**, Wiley and Sons.
18. a) Nozue, M. *Jpn. Kokai Tokkio Koho* Jp 63,211,251, **1988**. b) Nozue, M. *Jpn. Kokai Tokkio Koho* Jp O4,342,559, **1992**. c) Libman, M.B.; Shvets, V.F.; Suchov, Yu.P. *Khim.Prom.-st.,* **1988**, *9, 520.*
19. Kiyoura, T. *Jpn. Kokai Tokkio Koho* Jp 79,132,519, **1979**.
20. Oku, T.; Onda, Y.; Tsuneki, H.; Sumino, Y. *Jpn. Kokai Tokkio Koho* JP 07,112,953, **1995**.
21. Lanfranchi, M.; Prati, L.; Rossi, M.; Tiripicchio, A. *Journal of Chemical Society Chemical Communication*, **1993**, 1698.
22. a) Tsujino, T.; Ohigashi, S.; Sugiyama, S.; Kawashima, K.; Hayashi H. *Journal of Molecular Catalysis*, **1992**, *71*, 25. b) Hayashi, H. *Journal of Molecular Catalysis*, **1992**, 71, 25.
23. Behr, A.; Botulinski, A.; Carduck, F.J.; Schneider, M. *U.S. Patent* 5,321,156, **1994**.

11 Dimethylcarbonate: an answer to the need for safe chemicals

F. Rivetti

Summary

Implementation of green chemistry is based on the development of alternative safe processes and products. In this area dimethylcarbonate (DMC) plays a significant role both as a final product and as a versatile chemical intermediate. A 'from the cradle to the grave' analysis shows that features of DMC and related DMC based processes fully match the concepts of a benign chemistry.

11.1 Introduction

The adverse effects and risks toward the environment and human health deriving from the production and use of many chemicals have become a matter of increasing social concern. As a consequence, more and more attention has been focused on the use of safer chemicals, through the proper design of clean processes and products. This approach is largely referred to as 'green chemistry'.[1]

As for production processes, green chemistry makes large use of concepts such as eliminating the use of toxic raw materials and intermediates; reducing the quantity and toxicity of all emissions and wastes leaving the process; reducing or eliminating solvents; allowing by-products to be recycled. The use of catalysts is often the key to milder reaction conditions and increased selectivity, which results in lower by-product formation and minimization of energy consumption.

As far as products are concerned, green chemistry strategy focuses on reducing impacts along the entire life cycle of the product, from the production process (including considerations of raw materials extraction or generation) to the ultimate disposal of the product.

For more than 15 years, EniChem has been a pioneer in the development of large-scale industrial production and uses of dimethylcarbonate (DMC).[2-7] The scope of this chapter is to illustrate the excellent match between the features of industrial DMC exploitation and the concepts of a benign chemistry, through a 'from the cradle to the grave' analysis focused on DMC production, properties, relevant DMC based processes, and disposal. The interest in DMC production and uses and the role it plays in the current chemistry scene is witnessed by the steeply and steadily growing number of relevant Chemical Abstracts citations in the last few years (Fig. 11.1).

11.2 Dimethylcarbonate production

Today dimethylcarbonate is mainly produced by methanol oxycarbonylation (Eq. 1).

$$CO + 2CH_3OH + \tfrac{1}{2}O_2 \rightarrow (CH_3O)_2CO + H_2O \qquad (1)$$

Two oxycarbonylation technologies are currently carried out on industrial scale. EniChem first developed in Italy a DMC production process based on a liquid phase one-step oxidative carbonylation of methanol in the presence of copper chlorides, using oxygen as a direct oxidant: methanol and a gaseous stream containing oxygen and carbon monoxide are fed under pressure to the reactor.[8-10] This process, industrially exploited since 1983, accounts today for about 20 000 tons year^{-1} DMC currently produced worldwide, further 100 000 tons year^{-1} production being scheduled to go on stream in the near future. Later, in 1993, a gas-phase two-steps technology based on palladium catalyzed carbonylation of methylnitrite intermediate was developed by Ube in Japan.[11]

The oxidative carbonylation process avoids the traditional use of phosgene as raw material (Eq. 2).

$$COCl_2 + 2CH_3OH \rightarrow (CH_3O)_2CO + 2HCl \qquad (2)$$

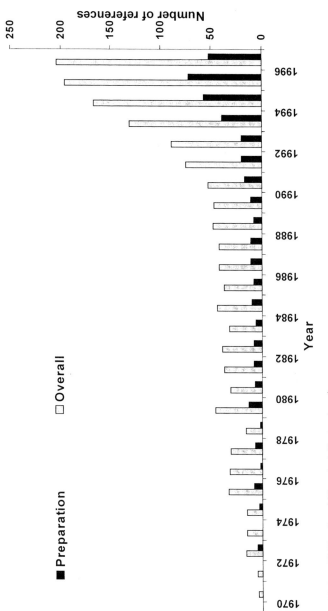

Fig. 11.1 DMC related Chemical Abstracts references.

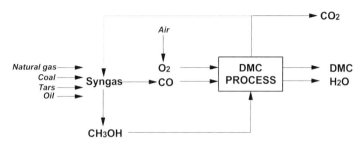

Fig. 11.2 DMC production by methanol oxycarbonylation.

Besides using highly toxic phosgene, the traditional process brings about coproduction of HCl, or chloride salts, and low purity of the product, which is contaminated by toxic, corrosive chlorinated impurities. As a consequence, historically, DMC used to be classified in Europe as an harmful chemical, whereas a rating outside any class of toxicity has now been acknowledged by the European Economic Community.[12]

On the contrary, the EniChem oxidative carbonylation process shows a number of distinctive favorable features, which turned DMC production into a clean technology (Fig. 11.2). Use of cuprous chloride as catalyst for the reaction allows high selectivity based on the reacted methanol, resulting in minimization of by-products and affording high purity of the product. Catalyst life is practically endless and, by proper process design, losses can be completely avoided, so that there is no need for catalyst disposal. The only coproducts are water and CO_2, which is formed from CO in substantial amount in the presence of water generated by the reaction. It is important to observe that the coproduced CO_2 can be efficiently reutilized as a carbon source in available CO generation processes. Moreover, as a raw material, CO can be produced by a variety of processes starting, besides oil derived feedstocks, from natural gas, coal, etc.

11.3 DMC properties

Available toxicity properties show that DMC has to be considered a very safe chemical both from the point of views of toxicity and ecotoxicity (Table 11.1). Acute and subchronic toxicity data on rats rank among the

Table 11.1 DMC toxicological and ecotoxicological properties.

Acute toxicity	LD_{50} (Oral, Rat) LD_{50} (Skin Rat) LC_{50} (Inh., Rat, 4h)		13.8 g kg^{-1} > 2.5g kg^{-1} 140 g m^{-3}
Subchronic toxicity	NOEL (Oral, Rat, 13 Weeks)	OECD 408	500 mg kg^{-1} d^{-1}
Mutagenic properties	Ames test DNA repair In vitro cytogenetic test In vitro cell gene mutation	OECD 473 OECD 473	Negative Negative Negative Negative
Irritation properties	Skin, Rabbit Skin, Rat Eye, Rabbit		Not irritant Not irritant Slight irritant
Aerobic biodegradation	MOD. MITI (28dd)	OECD 301C	90%
Acute toxicity on fish	NOEC (Golden Orfe)	OECD 203	1000 mg l^{-1}
Acute toxicity on aerobic waste water bacteria	EC_{50}	OECD 209	> 100 mg l^{-1}

safest values for chemicals. No evidence of mutagenetic as well as irritation properties were found, based on the performed tests. Its characteristics leave DMC outside any class of toxicity, the only classification of dangerousness required by EC legislation refers to its flammability properties. As for ecotoxicological properties, no indication of aquatic harm was observed, as judged from acute toxicity data on fish and aerobic waste water bacteria. Moreover, the product was found to be highly biodegradable. Tropospheric ozone formation potential, as a result of photochemical oxidation reactions in levels of significant nitrogen oxide pollution, is a topic of increasing concern nowadays. Recently, the atmospheric chemistry of DMC has been thoroughly investigated.[13] As a conclusion, the photochemical ozone forming potential of DMC was found to be negligible, when compared to other VOC, such as conventional fuels, due to fairly low reactivity towards OH radicals, comparable to that of ethane. On the other hand, the atmospheric lifetime of DMC has been calculated to be approximately 2 months, based on a reactivity 46 times higher than methylchloroform, suggesting that any risk of atmospheric accumulation is avoided.

11.4 DMC based processes

DMC based processes are characterized by a series of favorable aspects, fitting the guidelines for a clean chemistry:

1. Use of a harmless reagent: DMC is a substitute for toxic reactants, such as phosgene and chloroformates in carbonylations and alkoxycarbonylation reactions and methyl halides or dimethylsulphate in methylation reactions. More generally, alternative synthetic pathways exploiting DMC as an intermediate can be adopted, for example in the pharmaceutical field, shifting synthetic strategies to cleaner processes.

2. Absence of solvents: As a rule, reactions involving DMC are carried out by using an excess of DMC itself as a solvent, avoiding the further presence of process chemicals. This turns into simpler processes, easier separation procedures, reduction in energy consumption and lower probability of losses.

3. No salts or HCl coproduction: The use of traditional reactants, such as phosgene or methyl halides or sulphate, brings about, in the majority of cases, coproduction of salts, as a consequence of the stoichiometric consumption of NaOH or similar bases. These salts are

usually discharged in a waste water stream, contaminated by organics and requiring water treatment. In other cases, free HCl is released by the reaction and must be properly removed, for recovery or disposal. On the contrary, the majority of DMC based reactions requires only catalytic amounts of base catalyst, so that salt coproduction is negligible.

4. Easy disposable, recyclable by-products: As a result of methylation and carbonylation reactions by DMC, only CO_2 and methanol are released. Valuable methanol can be recycled as solvent or to other chemical uses; CO_2, at these production levels, does not involve disposal problems. If appropriate, recycle of methanol and even of a substantial part of CO_2 to DMC production facility is a real possibility.

5. Use of catalysts: DMC is a relatively stable, fairly unreactive chemical. Therefore, its exploitation as a chemical intermediate, substituting very reactive chemicals, normally requires the use of proper catalysts, able to trigger its reactivity. The use of catalysts is widely acknowledged as a key to clean chemistry. It helps to get better reactivity and selectivity control, increasing safety and reducing by-products.

Based on favorable features and economics, a number of industrial processes using DMC have been set up in the past years (Fig. 11.3). Applications, relying on both methylation and carbonylation reactions, rank from fine chemistry (pharmaceutical and veterinary products, agrochemicals, flavors and fragrances and electronics chemicals are some examples) to lubricants and polymers.

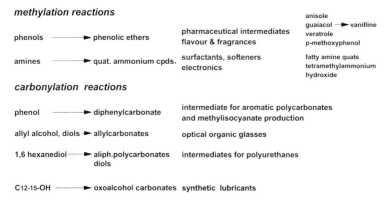

Fig. 11.3 DMC based processes: areas of application.

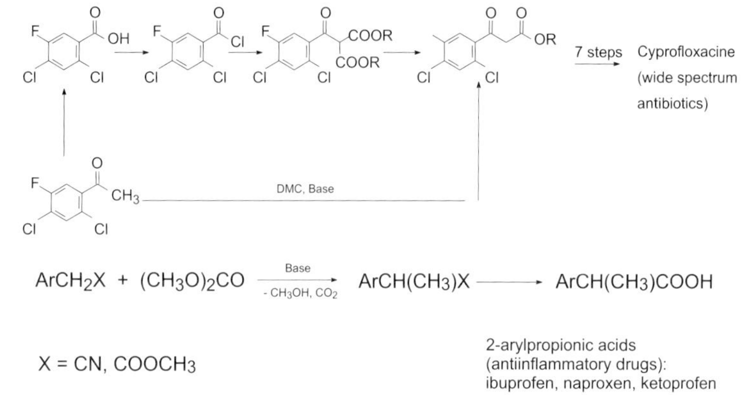

Fig. 11.4 Alternative DMC based pharmaceutical synthetic strategies.

Alternative pathways in synthetic strategies based on the use of DMC are well illustrated by two examples taken from the pharmaceutical field, selected for their actual or potential industrial interest (Fig. 11.4). The first one, currently industrially exploited, involves the production of the wide-spectrum antibiotic cyprofloxacine.[14] In the synthetic strategy to cyprofloxacine, C-carboalkoxylation by DMC at α-carbon in a substituted acetophenone allows the avoidance of three more synthetic steps, as traditionally performed. In the same way, the production of popular non-steroidal anti-inflammatory drugs, such as naproxen, ibuprofen and ketoprofen, chemically substituted arylpropionic acids, takes advantage from the use of DMC in a selective monomethylation step at the α-carbon of substituted benzyl esters or nitriles, as demonstrated up to a pilot scale.[15]

11.4.1 Aromatic polycarbonates production

By far the most prominent example of DMC industrial exploitation as a chemical intermediate is currently represented by the production of aromatic polycarbonates, in view of their great and still growing, industrial importance. The market for aromatic polycarbonates, including copolymers and polymer alloys, is estimated around 1 million tons year^{-1} and is further expanding. The DMC based route to aromatic polycarbonates, through the production of intermediate diphenylcarbonate (DPC) and melt polymerization between DPC and Bisphenol A

(BPA), is thouroughly challenging the previous technology based on interfacial polymerization, avoiding, at the same time, the use of phosgene as a reactant and methylene chloride as a solvent (Fig. 11.5).[16,17]

In interfacial polymerization process, BPA is transformed by NaOH to its sodium salt. A water solution of the salt is contacted with a CH_2Cl_2 solution of phosgene. This process requires large volumes of CH_2Cl_2 solvent and a complex finishing technology, to recover the polymer free of halogen impurities and catalyst residues, which is harmful to the following polymer processing by extrusion or molding (temperatures up to 340 °C are required). Chlorinated impurities are responsible for equipment corrosion and polymer yellowing and bring bad mechanical and weathering performances. In particular, the exhaustive removal of CH_2Cl_2 from the polymer is a hard task requiring special know-how. Also, the use of CH_2Cl_2 raises environmental problems, as a consequence of its release both in the atmosphere and in water waste streams, due to high volatility and appreciable water solubility. All these problems are overcome by the use of a melt polymerization process, which is carried out without phosgene and in the absence of any solvent (Fig. 11.6).

In the melt polymerization process, BPA and DPC obtained from DMC are transesterified to the polymer in the presence of a suitable catalyst, with removal of phenol, under increasingly higher vacuum and temperature conditions, for example up to 280 °C and 0.01 kPa. The melt polymerization process is characterized by high productivity, since it is operated in bulk, and it avoids the solvent recovery and waste treatment sections.

On the other hand, high vacuum and temperatures are needed for exhaustive phenol removal from the high viscosity melt, to reach the required molecular weight. Good know-how in catalyst choice and reactor design and in high vacuum technology is essential to obtain a good quality polymer without yellowing and structural defects. Another key success factor is the quality of the raw materials. Qualified DPC is best obtained by the transesterification process between DMC and phenol, whereas DPC produced from phosgene and sodium phenoxide, besides raising the problems associated with the use of phosgene in the process, normally contains chlorinated impurities and must be subjected to costly purification treatments.

Fig. 11.5 Aromatic polycarbonates: interfacial polymerization.

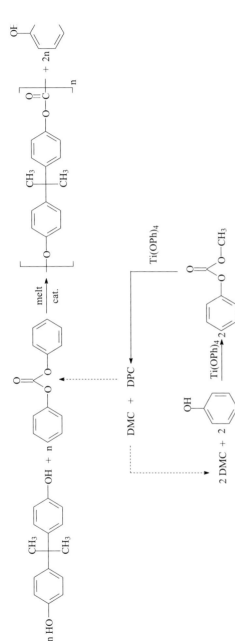

Fig. 11.6 Aromatic polycarbonates: melt polymerization.

According to the process developed by EniChem, the transesterification reaction to DPC is carried out in two steps, to overcome the particularly unfavorable thermodynamics of the reaction (Fig. 11.6).[4] In the first step, DMC and phenol are reacted in the presence of titanium alkoxide catalyst to give methyl phenyl carbonate (PMC). In the second step, PMC is disproportionated to DPC and DMC, which is recycled back to PMC production. As a result of this process scheme and proper engineering design, in spite of a largely positive free energy value (ca. +12.4 kcal mol^{-1}), the reaction is successfully carried out even on a quite large industrial scale.

In an overall picture, an integrated process can be designed from CO and O_2 to PC via DMC and DPC, with methanol and phenol internal recycles, resulting in no coproducts formation (Fig. 11.7).

11.4.2 Isocyanates production

Applications of DMC reactivity are far from being fully explored and exploited. DMC based, non-phosgene production of isocyanates would be of outstanding importance. Non-phosgene routes to isocyanates have been extensively pursued by several companies, mainly through thermolysis of the corresponding carbamate precursor (Fig. 11.8). The carbamate synthesis may involve a number of possible ways, such as: reaction of the nitrocompound with CO; or reaction of the amine with CO and O_2, urea and alcohol, or DMC.

The amine-DMC reaction has been proven effective in obtaining both aliphatic and aromatic carbamates. Mono and dicarbamates can be obtained as well, opening the road to a variety of commercial isocyanates. For example, hexamethylenediamine (HDA) is quickly transformed to the corresponding bis(methylcarbamate) in 99% yield by reaction with DMC at 65 °C in the presence of sodium methoxide as a catalyst. A following non-catalytic, gas-phase thermolysis step, carried out under vacuum (1.3 kPa) in the temperature range 420–460 °C, affords the diisocyanate (HDI) with 95% overall process yield based on the amine (Fig. 11.9).[18]

11.5 Further DMC applications

Besides the exploitation of its reactivity as a chemical intermediate, DMC behaves as a strong competitor in the field of solvents

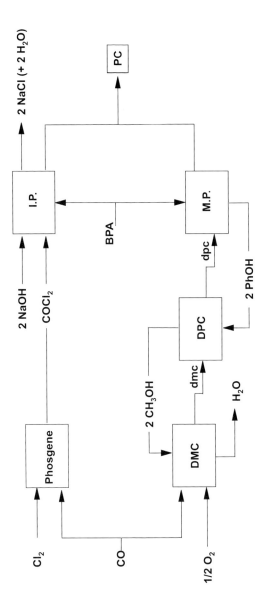

Fig. 11.7 DMC integrated melt polymerization process.

R-NO2, CO, CH3OH
-CO2
R-NH2, CO, O2, CH3OH
- H2O
R-NH2, CO(NH2)2, CH3OH
- NH3
R-NH2, DMC
- CH3OH
R-(NHCOOCH3)x
-x CH3OH
R-(NCO)x

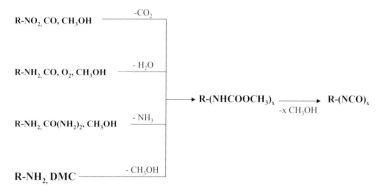

Fig. 11.8 Non-phosgene isocyanates routes.

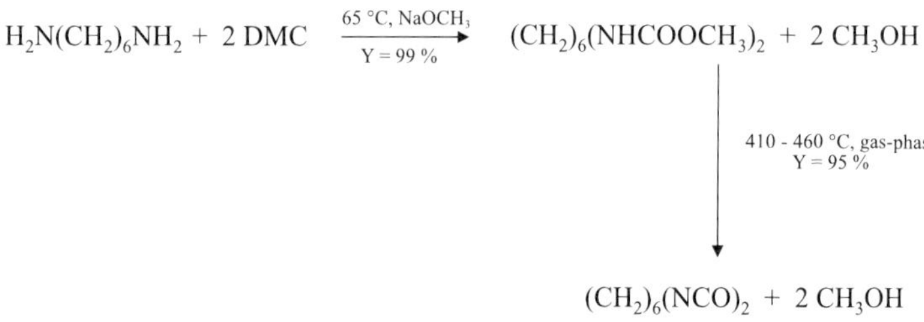

H2N(CH2)6NH2 + 2 DMC
65 °C, NaOCH3
Y = 99 %
(CH2)6(NHCOOCH3)2 + 2 CH3OH
410 - 460 °C, gas-phase
Y = 95 %
(CH2)6(NCO)2 + 2 CH3OH

Fig. 11.9 Hexamethylenediisocyanate *via* DMC.

(Table 11.2). It represents a viable alternative to acetate esters and ketones in most applications, from paints to adhesives, taking advantage of its good solvency power, for example toward commercial resins, and of its favorable characteristics, as far as human health and the environment are concerned.

It must also be remembered that DMC is the leader of an entire family of derived carbonic esters, easily available by straightforward transesterification reactions and mostly characterized by low toxicity and high biodegradability, whose properties, according to the targeted specific application, for example in the field of solvents, lubricants and plasticizers, can be tailored by selecting the proper molecular formula (Fig. 11.10).[19,20]

Examples highlighting how DMC and derivatives can help to manage technological and environmental problems, switching to the use of safer

Table 11.2 DMC as solvent.

Properties		DMC	EtOAc	BuOAc	MEK	MiBK
Viscosity	mPa.s, 20 °C	0.62	0.45	0.74	0.4	0.61
Boiling point	°C	90	78	126	80	117
Dielectric constant	25 °C	3.1	6.0	5.1	18.5	13.1
Solubility parameter	(Cal cm^{-3})$^{1/2}$	9.9	9.1	8.5	9.3	8.4
Water miscibility	Solvent water, %$_w$	12.8	7.7	0.7	26	2.0
	Water in solvent, %$_w$	3.3	3.3	1.2	12	2.4
Azeotrope with water	Boiling point, °C	78	71	91	73	88
	Solvent, %$_w$	87	92	73	89	76
Flash point	Closed cup, °C	16	–4	23	–4	14
LD$_{50}$	Oral, rat, g kg^{-1}	13.8	5.6	14	3.4	2.08

DMC: *dimethylcarbonate*; EtOAc: *ethyl acetate*; BuOAc: *butyl acetate*; MEK: *butan-2-one*; MiBK: *4-methyl pentan-2-one*

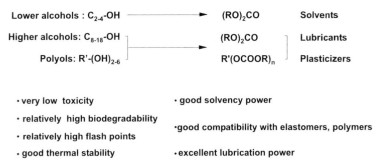

Fig. 11.10 Outlook of alkylcarbonate features.

chemicals, are represented by DMC application as blowing agent in the production of flexible PU foams, after CFC have been banned, overcoming alternative solutions such as the use of CH_2Cl_2 or cyclopentane;[21] and by the proposed formulation of biodegradable synthetic drilling muds for off-shore oil drilling, making use of higher carbonates.[22]

11.5.1 DMC and fuels

In recent years, DMC has been taken into consideration as a strong option for meeting compulsory oxygenate specification on gasoline in those countries such as the U.S. where local legislation requires a minimum oxygen content of between 2 and 2.7%.[23] The reason is the outstanding oxygen content in the DMC molecule, about three times the oxygen content of MTBE and 1.5 that of ethanol, which are the compounds currently mainly used for this purpose, leading to a much lower amount of DMC as oxygenate required to be added to the fuel (Table 11.3). The high oxygen content of DMC is coupled with a good blending octane number and other favorable blending properties (RVP, w/o distribution, etc.) and to the fact that the oxidative carbonylation production process allows cost-effective production, starting from a boundless, widely available raw material such as synthesis gas.

Although some debate still exists about the benefits, the use of oxygenated fuels is generally considered a proven and effective option to reduce vehicle emissions and associated environmental and health risks. Like other oxygenates, carbonates such as DMC have been found to be able to help reduce CO, HC and particulate emissions both from gasoline

Table 11.3 DMC as an oxygenate for fuels.

Compound	Oxygen content in the molecule, %$_w$	Volume for 2.7%$_w$ oxygen in gasoline, %
DMC	53.28	3.5
MTBE	18.15	15.1
Ethanol	34.73	7.7

DMC: *dimethylcarbonate*; MTBE: *methyl t-butyl ether*

and diesel vehicles, on the basis of added oxygen content to the fuel. Implementation of this opportunity will be linked to economic considerations and future legislative evolutions.

11.6 DMC disposal

In green chemistry the disposal of a chemical, at the end of its life cycle, should not cause any human or environmental damage. DMC is not expected to raise any problem from disposal, neither in water stream, by proper biological water treatment, due to its high aerobic biodegradability, nor by normal incineration, owing to the absence of heteroatoms besides oxygen in the molecule.

A chemical such as DMC, like many others, equates to a CO_2 emission in terms of ultimate disposal. From this point of view, it is worth noting that after combustion 1 kg DMC generates a much lower amount of CO_2 than many other common molecules (Table 11.4).

Acknowledgements

The author wishes to thank Enichem S.p.A. for authorising the publication of this work. He also thanks all the research group members and more specifically dr. Delledonne for their useful contribution.

Table 11.4 Comparative CO_2 emission from solvent disposal.

Compound	Emitted CO_2, kg kg^{-1}
DMC	1.4
Acetone	2.3
Ethyl acetate	2.0
Toluene	3.3

References

1. De Vito, S.C.; Garrett, R.L., eds; *Designing Safer Chemicals*; ACS Symp. Ser., 640, American Chemical Society: Washington DC, **1996**.
2. Cassar, L. *Chim. Ind. (Milan)* **1990**, *72*, 18–22.
3. Massi Mauri, M.; Romano, U. *Chimica Oggi* **1983**, *10*, 33–34.
4. Massi Mauri, M.; Romano, U.; Rivetti, F. *Quad. Ing. Chim. Ital.* **1985**, *21*, 6–12.
5. Rivetti, F.; Romano, U.; Delledonne, D. In *Green Chemistry: Designing Chemistry for the Environment*; ACS Symp. Ser., 626, Anastas, P.T., Williamson, T.C., eds; American Chemical Society: Washington DC, 1996; pp. 70–80.
6. Romano, U. *Chim. Ind. (Milan)* **1993**, *75*, 303–306.
7. Romano, U.; Rivetti, F. *Chimica Oggi* **1984**, *9*, 37–41.
8. Romano, U.; Tesei, R.; Massi Mauri, M.; Rebora, P. *Ind. Eng. Chem., Prod. Res. Dev.* **1980**, *19*, 396–403.
9. Romano, U.; Tesei R.; Cipriani, G.; Micucci, L. U.S. Pat. 4 218 391, 1980.
10. Di Muzio, N.; Fusi, C.; Rivetti, F.; Sasselli, G. U.S. Pat. 5 210 269, 1993.
11. Matsuzaki, T.; Nakamura, A. *Catalysis Surveys of Japan* **1997**, *1*, 77–88.
12. European Economic Community *Official Gazette, L 338A/62*, 1991.
13. Bilde M.; Møgelberg, T.E.; Sehested, J.; Nielsen, O.J.; Wallington T.J.; Hurtley M.D.; et al. *J. Phys. Chem. A*, **1997**, *101*, 3514–3525
14. Lopez Molina, I.; Palomo Coll, A.; Domingo Coto, A; et al. ES Pat. 2 009 072, 1988.
15. Tundo, P.; Selva, M. In *Green Chemistry: Designing Chemistry for the Environment*; ACS Symp. Ser. 626, Anastas, P.T., Williamson, T.C., eds; American Chemical Society: Washington DC, 1996; pp. 81–91.
16. Greco, A.; Rivetti, F. *Chim. Ind. (Milan)* **1998**, *80*, 77–83.
17. *Chemical Engineering* **1993**, *100 (5)*, 20.
18. Mizia, F.; Rivetti, F.; Romano, U. U.S. Pat. 5 315 034, **1994**.
19. Rivetti, F.; Romano, U. In *Proceedings of XXIst FATIPEC (Fédération d' Association Techniciens des Industries des Peintures, Vernis, Emaux d' Imprimerie de l' Europe Continental) Congress*, Amsterdam, **1992** Vol. 1, pp. 247–262.
20. Gerbaz, G.; Fisicaro, G. In *Synthetic lubricants and high performance functional fluids* R.L. Shubkin, eds; Dekker: New York, **1993** pp. 229–239.

21. Stefani, D.; Sam, F.O.; Lunardon, G. U.S. Pat. 5 340 845, **1994**.
22. Müller, H.; Herold, C.P; Westfechtel, A.; Von Tapavicze, S. DE Pat. 4 018 228, **1990**.
23. Pacheco, M.A.; Marshall, C.L. *Energy and Fuels*, **1997**, *11*, 2–29.

12 Environmentally benign organic transformations using microwave irradiation under solvent-free conditions

Rajender S. Varma

Introduction

Preventing pollution and minimizing waste generation to replace end-of-the-pipe control technology is an important aspect of green chemistry where the underlying theme emphasizes waste reduction 'at-source'. To achieve these preventive goals, the introduction of easily adaptable new processes or viable modifications of existing ones that reduce waste production is an attractive target. Microwave (MW) irradiation coupled with the use of recyclable mineral supports under solvent-free conditions is a unique approach that can prove beneficial in the development of eco-friendly synthetic protocols. A microwave approach has been used for a variety of applications including organic synthesis.[1–30] Such reactions involve selective absorption of MW energy by polar molecules, non-polar molecules being inert to MW dielectric loss. However, in these solution-phase reactions, the development of high pressures, and the use of specialized Teflon vessels are some of the limitations. Recently, we[3–29] and others[30] have added a practical dimension to the microwave heating protocols by accomplishing reactions under solvent-free conditions.[31] In these reactions, organic compounds adsorbed on the surface of inorganic oxides, such as alumina, silica and clay, or 'doped' supports absorb microwaves whereas the solid support does not absorb or restrict their transmission. These solvent-free MW-assisted reactions provide an opportunity to work with open vessels thus avoiding the risk of high pressure development and increasing the potential of such reactions to upscale. The solvent conservation made possible by these reactions can have an enormous impact on reducing waste discharge since solvents are often used in quantities 50–100 times those of the reacting materials. Herein, we describe our results on an environmentally benign MW approach for the synthesis of a wide variety of industrially important compounds and intermediates namely, enones, imines, enamines, nitroalkenes, oxidized sulfur species and heterocycles which, when otherwise obtained by conventional procedures contribute to the burden of chemical pollution. Thus, the problems associated with waste disposal of solvents (used several fold in chemical reactions) and excess chemicals are avoided or minimized. All the reactions described herein are performed in open glass containers (test tubes, beakers and round-bottomed flasks) using neat reactants under solvent-free conditions in an unmodified household MW oven operating at 2450 MHz (800 and 900 Watts). The comparisons of the MW-accelerated reactions have been made by conducting the same reaction in an oil bath at the same bulk temperature. Some of the supported reagents clay-supported iron(III) nitrate (clayfen) and copper(II) nitrate (claycop) are prepared according to the literature procedure.[32]

Methods, results and discussion

The practical feasibility has been demonstrated by conducting useful deprotection (cleavage) reactions of protected organic functional groups on inorganic surfaces as well as some condensation, oxidation and reduction reactions using MW irradiation. The general procedure involves simple mixing of neat reactants with the catalyst or their adsorption on to mineral or 'doped' supports and exposing the mixture to microwave irradion in an unmodified household MW oven.

12.1 Deprotection (cleavage) reactions

The protection–deprotection reaction sequences form an integral part of the preparation of monomer building blocks and fine chemicals and

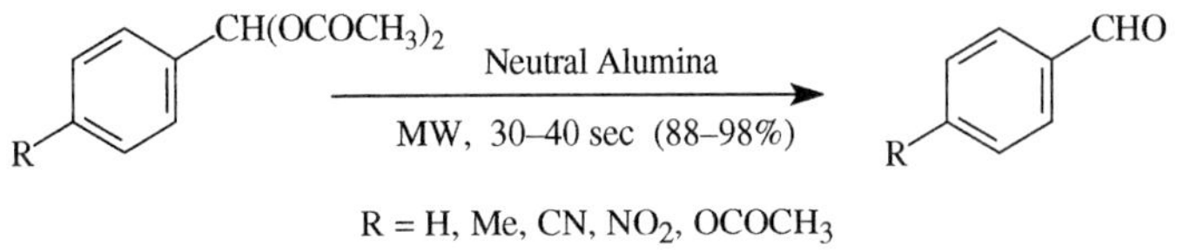

$$\text{R}\text{-}\text{C}_6\text{H}_4\text{-}\text{CH(OCOCH}_3)_2 \xrightarrow[\text{MW, 30–40 sec (88–98\%)}]{\text{Neutral Alumina}} \text{R}\text{-}\text{C}_6\text{H}_4\text{-}\text{CHO}$$

R = H, Me, CN, NO$_2$, OCOCH$_3$

Fig. 12.1 Cleavage of diacetrate derivatives of aromatic aldehydes.

precursors for pharmaceuticals and are often carried out using acidic, basic or hazardous and corrosive reagents and toxic metal salts. The solventless MW-assisted deprotection of protected alcohols and phenols,[6] aldehydes,[5,7] ketones,[4,10,11] including desilylation[8] and debenzylation[9] reactions have been successfully demonstrated.

12.1.1 Deacylation of benzaldehyde diacetates

The diacetate derivatives of aromatic aldehydes are rapidly cleaved on neutral alumina surface upon exposure to MW irradiation (Fig. 12.1). The selectivity in these deprotection reactions is demonstrated by simply adjusting the time of irradiation. As an example, for molecules bearing acetoxy functionality (R=OCOCH$_3$), the aldehyde diacetate is selectively removed in 30 sec, whereas an extended period of 2 min is required to cleave both the diacetate and ester groups. The yields obtained are better than those possible by conventional methods and the protocol is applicable to compounds encompassing olefinic moieties such as cinnamaldehyde diacetate.

12.1.2 Debenzylation of carboxylic esters

The promising solvent-free debenzylation of esters (Fig. 12.2)[9] paves the way for the cleavage of the 9-fluorenylmethoxycarbonyl (Fmoc) group that can be extended to protected amines by changing the surface characteristics of the solid support. The optimum conditions for cleavage of N-protected moieties require the use of basic alumina and irradiation time of 12–13 min time at ~130–140 °C. This approach, which may find application in peptide bond formation, would eliminate the use of irritating and corrosive chemicals such as trifluoroacetic acid and piperidine.

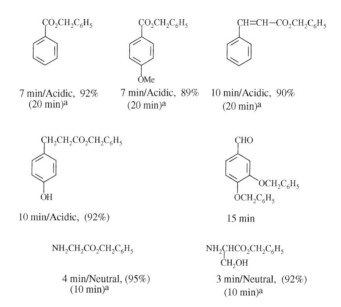

Fig. 12.2 Cleavage of benzyl esters via microwave thermolysis on alumina.
a) Time in parentheses refer to deprotection in oil bath at the same temperature.

12.1.3 Desilylation and deacylation of protected alcohols

t-Butyldimethyl silyl (TBDMS) ether derivatives of a variety of alcohols are rapidly deprotected to regenerate the corresponding hydroxy compounds on alumina surface under MW irradiation conditions (Fig. 12.3).[8] This approach eliminates the use of corrosive fluoride ions which are conventionally used in cleaving the silyl protecting groups.[8]

12.1.4 Deacylation reactions

The orthogonal deprotection of alcohols is possible on neutral alumina surface using microwave irradation (Fig. 12.4). Interestingly, chemoselectivity between alcoholic and phenolic groups in the same molecule can be achieved simply by varying the reaction time; the phenolic acetates are deacetylated faster than alcoholic analogues.[6]

The optimization of relevant parameters such as the power level of microwave employed and pulsed techniques (multi-stage, discontinuous irradiation to avoid the generation of higher temperatures) can be used to obtain the desired results.

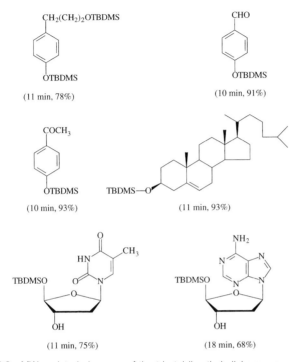

Fig. 12.3 MW-assisted cleavage of the t-butyldimethyl silyl groups.

12.1.5 Regeneration of carbonyl compounds
12.1.5.1 Dethioacetalization

Among the regeneration of carbonyl compounds, the cleavage of acid and base stable thioacetals and thioketals is quite challenging; the deprotection of thioacetals invariably requires the use of toxic heavy metals such as Hg^{+2}, Ag^{+2}, Ti^{+4}, Cd^{+2}, Tl^{+3}, or reagents such as

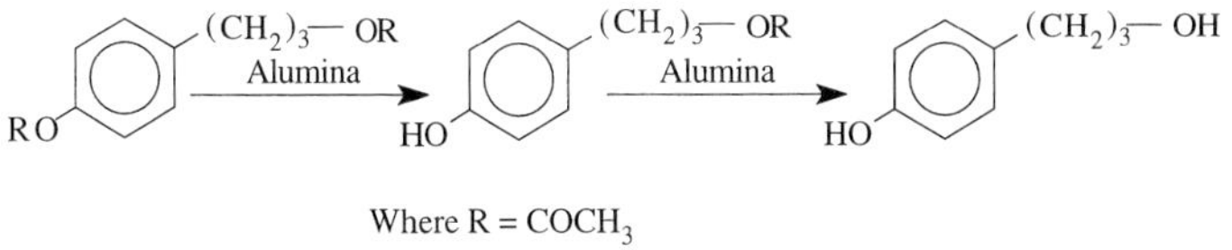

Fig. 12.4 Orthogonal deprotection of alcohols.

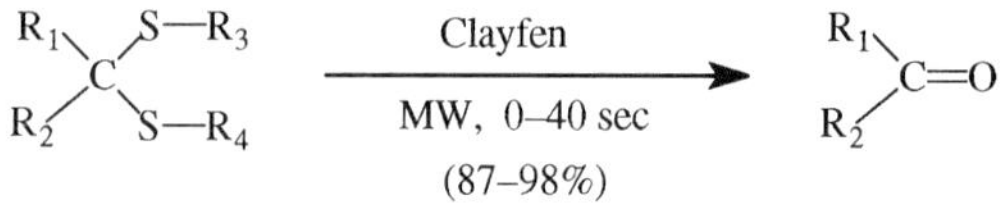

$$R_1R_2C(SR_3)(SR_4) \xrightarrow[\text{MW, 0–40 sec} \\ (87–98\%)]{\text{Clayfen}} R_1R_2C{=}O$$

$R_1 = R_2 = Et$; R_3–$R_4 = -(CH_2)_2-$; $R_1 = R_2 = Ph$; $R_3 = R_4 = Et$
$R_1 = Ph$, p-anisyl, p-$NO_2C_6H_4$: $R_2 = H$; R_3–$R_4 = -(CH_2)_2-$
R_1–R_2 = 2-Methylcyclohexyl, isoflavanolyl : R_3–$R_4 = -(CH_2)_2-$
$R_1 = Ph$: $R_2 = Me$: R_3–$R_4 = -(CH_2)_2-$

Fig. 12.5 Dethioacetalization.

benzeneseleninic anhydride.[5] This dethioacetalization reaction can be accomplished in high yield and in solid state using clayfen (Fig. 12.5).[5]

12.1.5.2 Deoximation reactions

The important role of oximes as protecting groups due to their hydrolytic stability has provided motivation for the development of newer deoximation reagents such as Raney nickel, pyridinium chlorochromate, pyridinium chlorochromate-H_2O_2, triethylammonium chlorochromate, dinitrogen tetroxide, trimethylsilyl chlorochromate, Dowex-50, dimethyl dioxirane, H_2O_2 over titanium silicalite-1, zirconium sulfophenyl phosphonate, N-haloamides and bismuth chloride.[4]

12.1.5.3 Using ammonium persulfate on silica

The deprotection of protected carbonyl compounds by ammonium persulfate on silica (Fig. 12.6) has been successfully demonstrated.[4] Neat oximes are admixed with the solid supported reagent and the contents are irradiated at full power in a MW oven to regenerate free aldehydes or ketones in a process that is applicable to both aldoximes and ketoximes. The role of surface is critical since the same reagent

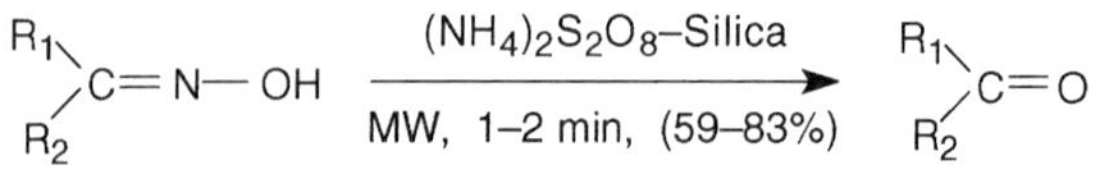

$$R_1R_2C{=}N{-}OH \xrightarrow[\text{MW, 1–2 min, (59–83\%)}]{(NH_4)_2S_2O_8-\text{Silica}} R_1R_2C{=}O$$

$R_1 = Ph$, 4-ClC_6H_4, 4-MeC_6H_4, 4-$MeOC_6H_4$; $R_2 = CH_3$
$R_1 = Ph$, 4-$NO_2C_6H_4$, 3,4-$(MeO)_2C_6H_4$, 2-thienyl, 1-naphthyl ; $R_2 = H$
$R_1 = R_2 = $ cyclopentyl

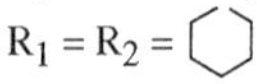

Fig. 12.6 Deprotection of protected carbonyl compounds.

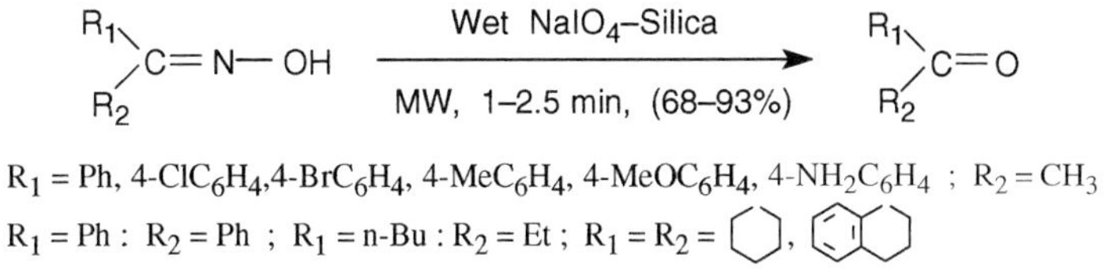

$$\underset{R_2}{\overset{R_1}{>}}C=N-OH \xrightarrow[\text{MW, 1–2.5 min, (68–93\%)}]{\text{Wet NaIO}_4\text{–Silica}} \underset{R_2}{\overset{R_1}{>}}C=O$$

R_1 = Ph, 4-ClC$_6$H$_4$, 4-BrC$_6$H$_4$, 4-MeC$_6$H$_4$, 4-MeOC$_6$H$_4$, 4-NH$_2$C$_6$H$_4$; R_2 = CH$_3$
R_1 = Ph : R_2 = Ph ; R_1 = n-Bu : R_2 = Et ; R_1 = R_2 = ⌬, ⌬⌬

Fig. 12.7 Deoximation.

supported on clay surface delivers predominantly the Beckmann rearranged products, the amides.[4]

12.1.5.4 Using NaIO$_4$ on silica

A very facile method for deoximation with sodium periodate on moist silica (Fig. 12.7) has been introduced that is applicable exclusively to ketoximes.[10]

12.1.5.5 Cleavage of semicarbazones and hydrazones

Aldehydes and ketones are also rapidly regenerated from the corresponding semicarbazones and hydrazones using ammonium persulfate impregnated on montmorillonite K 10 clay (Fig. 12.8) under both microwave or ultrasound irradiation conditions.[11] The results are documented in Table 12.1.

12.1.5.6 Dethiocarbonylation

Several reagents such as trifluoroacetic anhydride, CuCl/MeOH/NaOH, tetrabutylammonium hydrogen sulfate/NaOH, clay/ferric nitrate, NOBF$_4$, bromate and iodide solutions, alkaline hydrogen peroxide, sodium peroxide, bases for example KOBu, thiophosgene, DMSO, trimethyloxonium fluoborate, tellurium based oxidants, photochemical transformations, dimethyl selenoxide, benzeneseleninic anhydride, benzoyl peroxide, halogen-catalyzed alkoxides under phase transfer conditions, NaNO$_2$/HCl, Hg(OAc)$_2$, SOCl$_2$/CaCO$_3$, and singlet oxygen

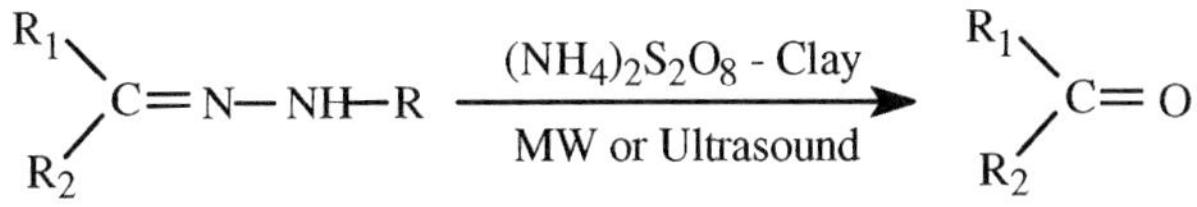

$$\underset{R_2}{\overset{R_1}{>}}C=N-NH-R \xrightarrow[\text{MW or Ultrasound}]{(NH_4)_2S_2O_8 \text{ - Clay}} \underset{R_2}{\overset{R_1}{>}}C=O$$

Fig. 12.8 Regeneration of aldehydes and Ketones.

Table 12.1 Cleavage of semicarbazones and phenylhydrazones with ammoniam persulfate and clay using microwave or ultrasonic irradiation

Entry	R_1	R_2	R	MW (Ultrasound) Time [min (h)]	Yield (%)
1	Ph	Me	$CONH_2$	0.6 (0.75)	72 (77)
2	p-Cl-C_6H_4	Me	$CONH_2$	1.8 (3.00)	82 (90)
3	p-HO-C_6H_4	Me	$CONH_2$	1.0 (1.00)	85 (94)
4	p-NH_2-C_6H_4	Me	$CONH_2$	1.0 (1.50)	65 (62)
5	p-MeO-$C_6H_4CH_2CH_2$	Me	$CONH_2$	1.5 (3.00)	75 (71)
6	n-Bu	Et	$CONH_2$	1.3 (2.50)	70 (79)
7	Ph	Me	Ph	0.5 (1.50)	71 (82)
8	n-Bu	Et	Ph	1.3 (1.00)	65 (71)
9	(cyclohexyl)		$CONH_2$	0.7 (1.50)	68 (69)
10	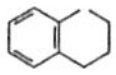 (tetralinyl)		$CONH_2$	1.0 (2.00)	69 (58)

Ratio of semicarbazones/phenylhydrazones to ammonium persulfate on clay is 1 : 8 (molar).

have been used for dethiocarbonylation.[12] But these methods have certain limitations such as the use of the stoichiometric amounts of the oxidants which are often inherently toxic or require longer reaction time or involve tedious procedures. A variety of thioketones are readily converted into the corresponding ketones under solvent-free conditions using clayfen or clayan in a process that is accelerated by microwave irradiation (Figs. 12.9 and 12.10).[12]

12.2 Oxidation reactions

12.2.1 Oxidation of alcohols

The conventional oxidizing reagents employed for organic functionalities are peracids, peroxides, manganese dioxide (MnO_2), potassium

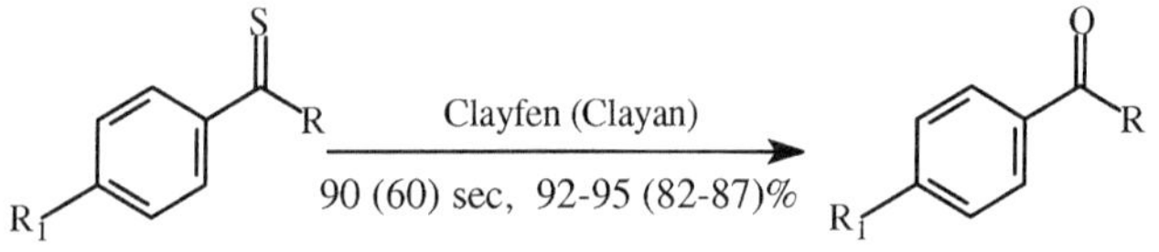

Clayfen (Clayan)
90 (60) sec, 92-95 (82-87)%

R = Me : R_1 = H ; R = Ph : R_1 = H, Br, Me

Fig. 12.9 Dethiocarbonylation.

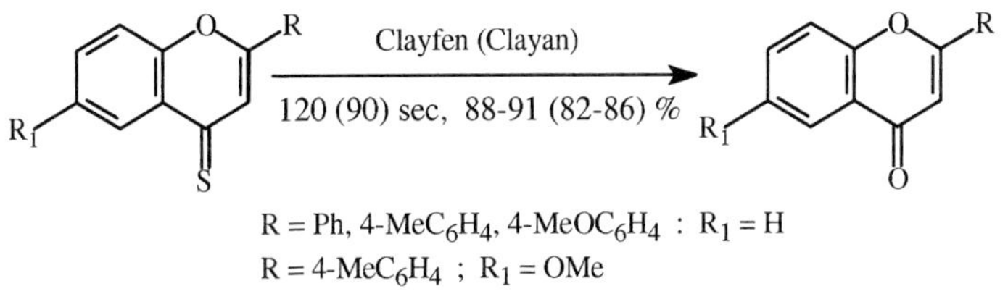

[Scheme: chromone-thione + Clayfen (Clayan), 120 (90) sec, 88–91 (82–86) % → chromone]

R = Ph, 4-MeC$_6$H$_4$, 4-MeOC$_6$H$_4$: R$_1$ = H
R = 4-MeC$_6$H$_4$; R$_1$ = OMe

Fig. 12.10 Dethiocarbonylation.

permanganate (KMnO$_4$), chromium trioxide (CrO$_3$), potassium chromate (K$_2$CrO$_4$) and potassium dichromate (K$_2$Cr$_2$O$_7$),[13] though these reagents have their own limitations in terms of toxicity, work-up and associated waste disposal problems.

Metal-based reagents have been extensively used in organic synthesis. The utility of such reagents in the oxidative transformation is compromised due to their inherent toxicity, cumbersome preparation, potential danger (ignition or explosion) in handling of its complexes, difficulties in terms of product isolation and waste disposal. Introduction of metallic reagents on solid supports has circumvented some of these problems and provided an attractive alternative in organic synthesis because of the selectivity and associated ease of manipulation. Further, the immobilization of metals on the surface avoids their leaching into the environment.

12.2.1.1 Selective and solventless oxidation with clayfen

A facile oxidation of alcohols to carbonyl compounds is reported here using montmorillonite K 10 clay-supported iron(III) nitrate (clayfen), under solvent-free conditions, in a process that is accelerated by MW irradiation.[13] The reaction presumably proceeds *via* the intermediacy of nitrosonium ions, and importantly no formation of carboxylic acids occurs in the oxidation of primary alcohols. The experimental procedure involves a simple mixing of neat alcohols with clayfen and irradiation of the reaction mixtures in a MW oven for 15–60 seconds in the absence of any solvent. This extremely rapid, manipulatively simple, inexpensive and selective protocol avoids the use of excess solvents and toxic oxidants. Using clayfen [iron(III) nitrate] in solid state and in amounts that are half that of used by Laszlo *et al.*,[32] we have achieved a rapid synthesis of carbonyl compounds in high yields (Fig. 12.11).[13]

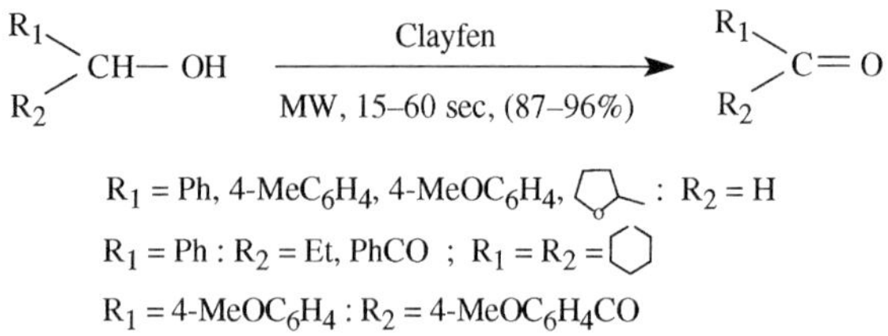

$$\underset{R_2}{\overset{R_1}{\diagdown}}CH-OH \xrightarrow[\text{MW, 15–60 sec, (87–96\%)}]{\text{Clayfen}} \underset{R_2}{\overset{R_1}{\diagdown}}C=O$$

R_1 = Ph, 4-MeC$_6$H$_4$, 4-MeOC$_6$H$_4$, (furyl) : R_2 = H

R_1 = Ph : R_2 = Et, PhCO ; R_1 = R_2 = (cyclohexyl)

R_1 = 4-MeOC$_6$H$_4$: R_2 = 4-MeOC$_6$H$_4$CO

Fig. 12.11 Oxidation of alcohols to carbonyl compounds.

12.2.1.2 Using activated manganese dioxide

Using manganese dioxide–silica, an expeditious and high yield route to carbonyl compounds is accomplished. Benzyl alcohols are selectively oxidized to carbonyl compounds using 35% MnO$_2$ 'doped' silica under MW irradiation conditions (Fig. 12.12).[14]

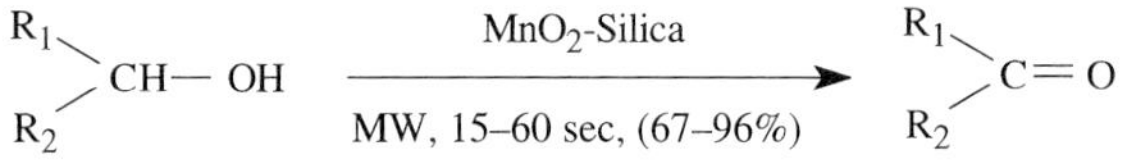

$$\underset{R_2}{\overset{R_1}{\diagdown}}CH-OH \xrightarrow[\text{MW, 15–60 sec, (67–96\%)}]{\text{MnO}_2\text{-Silica}} \underset{R_2}{\overset{R_1}{\diagdown}}C=O$$

R_1 = Ph, 4-MeC$_6$H$_4$, 4-MeOC$_6$H$_4$, PhCH=CH : R_2 = H

R_1 = Ph : R_2 = Et, Ph, PhCO ; R_1 = R_2 = hydroquinone

R_1 = 4-MeOC$_6$H$_4$: R_2 = 4-MeOC$_6$H$_4$CO

Fig. 12.12 Oxidation of alcohols using manganese dioxide–silica.

12.2.1.3 Using claycop–hydrogen peroxide

Metal ions play a significant role in many of these oxidative reactions as well as in biological dioxygen metabolism. Copper(II) acetate and H$_2$O$_2$ have been used to produce a stable oxidizing agent, hydroperoxy copper(II) compound, which is also obtainable from copper(II) nitrate and hydrogen peroxide (Eq. 1). The ensuing nitric acid requires neutralization by potassium bicarbonate to maintain a pH ~5.

$$2Cu(NO_3)_2 + H_2O_2 + 2H_2O \rightarrow 2CuO_2H + 4HNO_3 \qquad (1)$$

Copper(II) nitrate impregnated on K 10 clay (Claycop)–hydrogen peroxide system is an effective reagent for the oxidation of a variety of substrates and provides excellent yields (Fig. 12.13, Table 12.2).[15] The method does not require the maintenance of pH of the reaction mixture.

Environmentally benign organic transformations 231

Table 12.2 Oxidation of organic substrates by claycop and hydrogen peroxide

Entry	R_1	R_2	R_3	Time (min)	Yield (%)
1	Ph	H	Br	0.50	75
2	Ph	H	CN	0.75	80
3	Ph	H	NH_2	0.50	76
4	Ph	H	COOH	1.00	83
5	Ph	Ph	COOH	1.30	82
6	p-$NO_2C_6H_4$	H	H	1.30	69
7	Triphenylphosphine			0.25	85
8	Hydroquinone			0.50	71

The mole ratio of the substrate, copper(II) nitrate and hydrogen peroxide used was 1 : 0.8 : 1, respectively. Using 0.8 mmol equivalents of the copper(II) nitrate, phenyl acetonitrile was converted into phenylacetic acid.

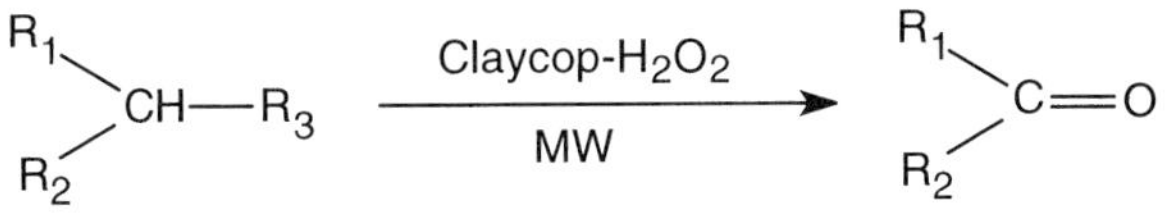

$$\underset{R_2}{\overset{R_1}{>}}CH-R_3 \quad \xrightarrow[MW]{Claycop-H_2O_2} \quad \underset{R_2}{\overset{R_1}{>}}C=O$$

Fig. 12.13 Oxidation using claycop–hydrogen peroxide.

12.2.1.4 Using chromium trioxide impregnated on wet alumina

The utility of chromium(VI) reagents in the oxidative transformation is compromised due to toxicity, involved preparation of its various complexes and cumbersome work-up and disposal problems. Chromium trioxide (CrO_3) impregnated on premoistened alumina is an efficient alternative system which oxidizes benzyl alcohols by simple admixing at room temperature (Fig. 12.14). The reactions are relatively clean with no

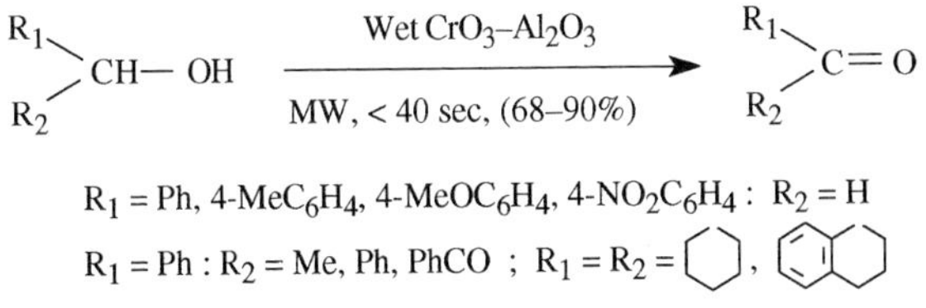

$$\underset{R_2}{\overset{R_1}{>}}CH-OH \quad \xrightarrow[MW, < 40 \text{ sec, } (68-90\%)]{\text{Wet } CrO_3-Al_2O_3} \quad \underset{R_2}{\overset{R_1}{>}}C=O$$

R_1 = Ph, 4-MeC_6H_4, 4-$MeOC_6H_4$, 4-$NO_2C_6H_4$: R_2 = H
R_1 = Ph : R_2 = Me, Ph, PhCO ; R_1 = R_2 = ⬡, ⬡⬡

Fig. 12.14 Oxidation using chromium trioxide on wet alumina.

tar formation, typical of many CrO_3 oxidations. Interestingly, no overoxidation to carboxylic acids is observed.[16]

12.2.1.5 Using iodobenzene diacetate on alumina

Iodoxybenzene, o-iodoxybenzoic acid (IBX), bis(trifluoroacetoxy)iodobenzene (BTI) and Dess–Martin periodinane are some of the common organohypervalent iodine reagents which have been used for the oxidation of alcohols and phenols but the use of iodobenzene diacetate (IBD) in this area, in spite of its low cost, has not been fully exploited. Also, IBX has been reported to be explosive under heavy impact and heating over 200 °C. Most of these reactions, however are conducted in high boiling DMSO and toxic acetonitrile media that results in environmental pollution.

A facile oxidation of alcohols to carbonyl compounds using alumina-supported IBD occurs rapidly under solvent-free conditions and MW irradiation in quantitative yields.[17] The advantage of using alumina as a support is apparent from the improved yields obtained using alumina–IBD system as compared to neat IBD (Fig. 12.15). Interestingly, 1,2-benzenedimethanol undergoes cyclization to afford 1(3H)-isobenzofuranone. The results are documented in Table 12.3.

Table 12.3 Oxidation of alcohols with IBD on alumina under MW irradiation

Entry	R_1	R_2	PhI(OAc)$_2$		Al$_2$O$_3$-PhI(OAc)$_2$	
			Time (min)	Yields (%)	Time (min)	Yields (%)
1	Ph	H	2.0	89	1.0	94
2	p-MeC$_6$H$_4$	H	2.0	88	1.0	92
3	p-MeOC$_6$H$_4$	H	1.5	91	2.0	95
4	Ph	Et	2.0	83	2.0	89
5	Ph	PhCO	2.0	90	2.0	90
6	1,2-benzenedimethanol		1.5	81	1.5	86
7	Anisoin		a	a	3.0	96
8	Hydroquinone		a	a	1.0	69
9	Catechol		a	a	0.5	43

a denotes a mixture of products.

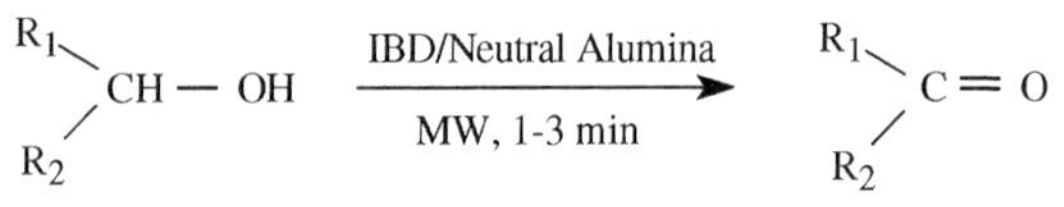

$$\underset{R_2}{\overset{R_1}{>}}CH-OH \xrightarrow[\text{MW, 1-3 min}]{\text{IBD/Neutral Alumina}} \underset{R_2}{\overset{R_1}{>}}C=O$$

Fig. 12.15 Oxidation of alcohols using alumina–IBD.

12.2.1.6 Using copper sulfate–alumina or oxone®–wet alumina

The oxidative transformation of benzoins to benzils has been accomplished by a variety of reagents namely nitric acid, Fehling's solution, thalium(III) nitrate (TTN), ytterbium(III) nitrate, clayfen and ammonium chlorochromate–alumina.[18] Besides the extended reaction time required, most of these processes suffer from drawbacks such as the use of corrosive acids and toxic metallic compounds that generate undesirable waste materials. Consequently, there is a need for the development of a manipulatively easy and an environmentally benign, solvent-free protocol for the oxidation of benzoins. We discovered that both the symmetrical and unsymmetrical benzoins are rapidly oxidized to benzils in high yields using the solid reagent systems, copper(II) sulfate–alumina[18] or Oxone®–wet alumina[19] under the influence of microwaves (Fig. 12.16).

Intrestingly, under these solvent-free reaction conditions, primary alcohols (for example benzyl alcohol) and secondary alcohols (for example 1-phenyl-1-propanol) undergo only limited oxidative conversion which is of little practical utility. Apparently, the process is applicable only to α-hydroxyketones as exemplified by various substrates including a mixed benzylic/aliphatic α-hydroxyketone, 2-hydroxypropiophenone that affords the corresponding vicinal diketone.[19]

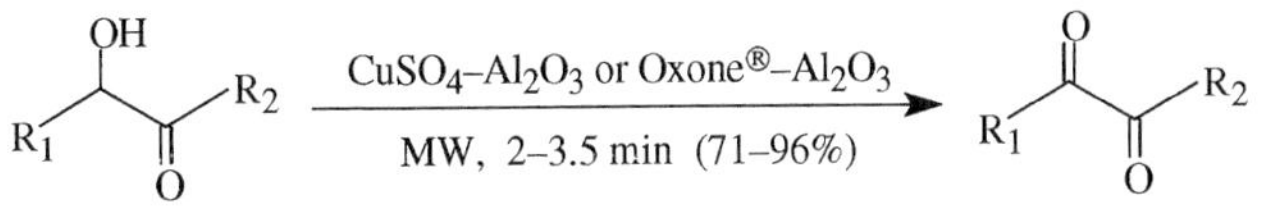

$$R_1\text{-CH(OH)-CO-}R_2 \xrightarrow[\text{MW, 2-3.5 min (71-96\%)}]{\text{CuSO}_4\text{-Al}_2\text{O}_3 \text{ or Oxone}^\circledR\text{-Al}_2\text{O}_3} R_1\text{-CO-CO-}R_2$$

Where $R_1 = R_2 = C_6H_5$, $p\text{-}CH_3C_6H_4$, $p\text{-}CH_3OC_6H_4$, $p\text{-}ClC_6H_4$, furyl

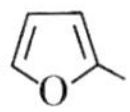

$R_1 = C_6H_5$; $R_2 = p\text{-}CH_3C_6H_4$, $p\text{-}CH_3OC_6H_4$

and $R_1 = CH_3$; $R_2 = C_6H_5$

Fig. 12.16 Oxidation using copper sulfate–alumina or oxone® – wet alumina.

12.2.2 Oxidation of sulfides

12.2.2.1 Using sodium periodate

Sulfides are usually oxidized to sulfoxides under drastic conditions using strong oxidants such as nitric acid, hydrogen peroxide, chromic acid, peracids and periodate.[20] Using MW irradiation this oxidative transformation is achievable with desired selectivity in oxidation to either sulfoxides or sulfones using silica 'doped' with 10% sodium periodate and by varying the power of irradiation and reaction time (pulsed techniques).[20] Consequently, a much reduced amount of the active oxidizing agent is employed that is safer to handle (Fig. 12.17).

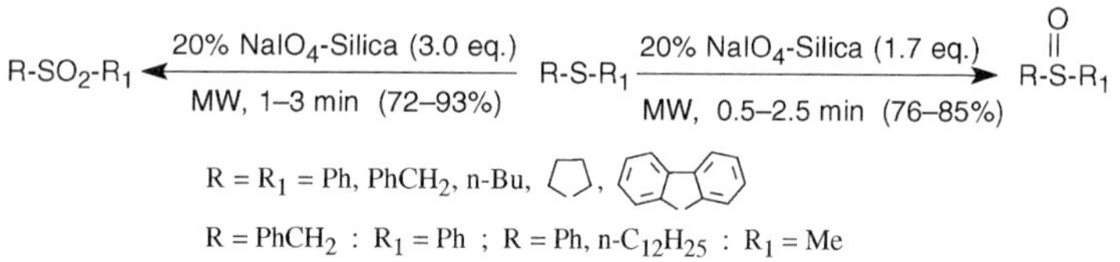

$$R\text{-}SO_2\text{-}R_1 \xleftarrow[\text{MW, 1-3 min (72-93\%)}]{20\% \text{ NaIO}_4\text{-Silica (3.0 eq.)}} R\text{-}S\text{-}R_1 \xrightarrow[\text{MW, 0.5-2.5 min (76-85\%)}]{20\% \text{ NaIO}_4\text{-Silica (1.7 eq.)}} R\text{-}\overset{\overset{O}{\|}}{S}\text{-}R_1$$

R = R$_1$ = Ph, PhCH$_2$, n-Bu, (cyclopentyl), (benzothiophene)

R = PhCH$_2$: R$_1$ = Ph ; R = Ph, n-C$_{12}$H$_{25}$: R$_1$ = Me

Fig. 12.17 Oxidation of sulfides using sodium periodate.

Importantly, various refractory thiophenes that are often not reductively removed by conventional refining processess can be oxidized under these conditions; for example benzothiophenes are oxidized in solid state to the corresponding sulfoxides and sulfones using ultrasonic and microwave irradiation, respectively, in the presence of NaIO$_4$–silica.[20] A noteworthy feature of the protocol is its applicability to long chain fatty sulfides which are insoluble in most solvents and are consequently difficult to oxidize.

12.2.2.2 Using iodobenzene diacetate on alumina

As described earlier, the solid reagent system, IBD–alumina, is a useful oxidizing agent and can be used for expeditious and selective oxidation of alkyl, aryl and cyclic sulfides to the corresponding sulfoxides in excellent yields upon MW activation (Fig. 12.18).[21]

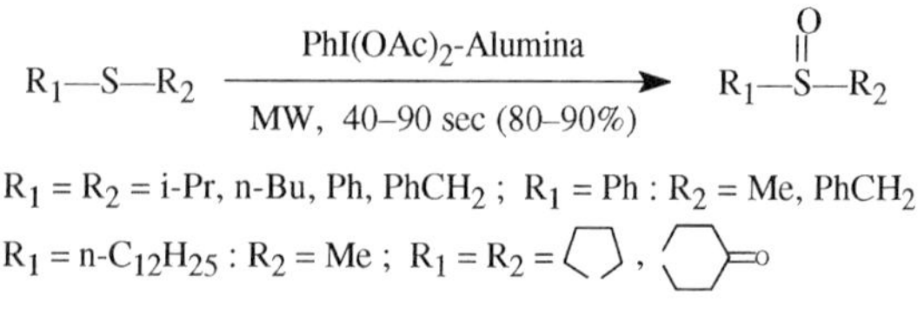

$$R_1\text{—}S\text{—}R_2 \xrightarrow[\text{MW, 40–90 sec (80–90\%)}]{\text{PhI(OAc)}_2\text{-Alumina}} R_1\text{—}\overset{\overset{O}{\|}}{S}\text{—}R_2$$

R$_1$ = R$_2$ = i-Pr, n-Bu, Ph, PhCH$_2$; R$_1$ = Ph : R$_2$ = Me, PhCH$_2$

R$_1$ = n-C$_{12}$H$_{25}$: R$_2$ = Me ; R$_1$ = R$_2$ = (cyclopentyl), (cyclohexanone)

Fig. 12.18 Oxidation of sulfides using IBD–alumina.

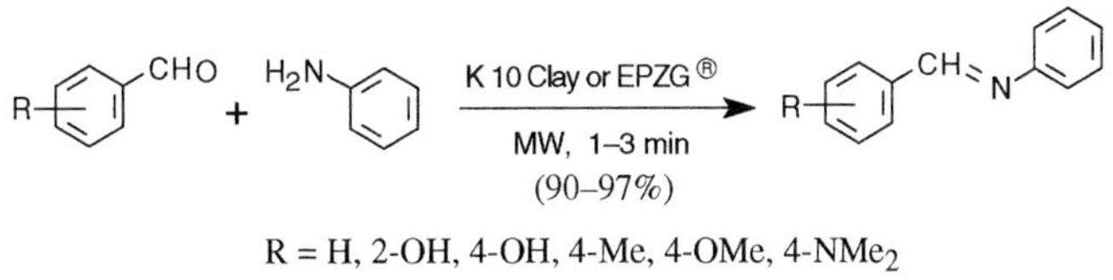

Fig. 12.19 Synthesis of imines.

12.3 Condensation reactions

12.3.1 Synthesis of imines, enamines and nitroalkenes

The azeotropic removal of water from the intermediate is the driving force for the preparation of imines, enamines and nitroalkenes in reactions that are normally catalyzed by *p*-toluenesulphonic acid, titanium(IV) chloride and montmorillonite K 10 clay. Conventionally, a Dean Stark's apparatus is used which requires a large excess of aromatic hydrocarbons such as benzene or toluene for azeotropic water elimination.

MW-induced acceleration of such dehydration reactions using a catalytic montmorillonite K 10 clay[22] (Figs 12.19 and 12.20) or Envirocat reagent,[23] EPZG® (Figs 12.19 and 12.20) has been demonstrated in a facile preparation of imines and enamines via the reactions of primary and secondary amines with aldehydes and ketones, respectively. Microwaves, generated at the common frequency of 2450 MHz, are ideally suited to remove water in imine- or enamine-forming reactions. For low boiling reactants, deployment of variable power intensities of microwaves coupled with pulsed techniques have been used.

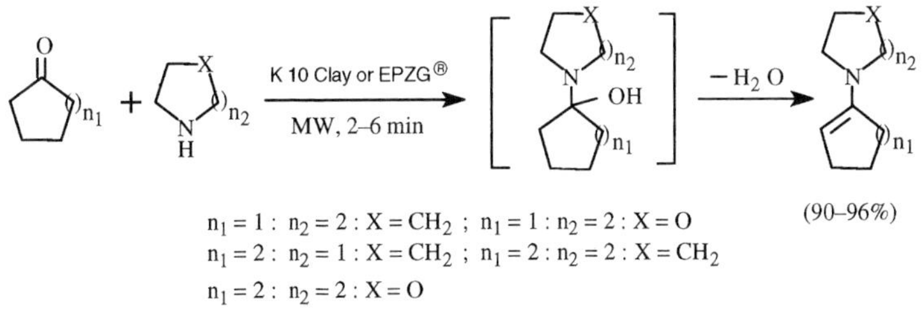

Fig. 12.20 Synthesis of enamines.

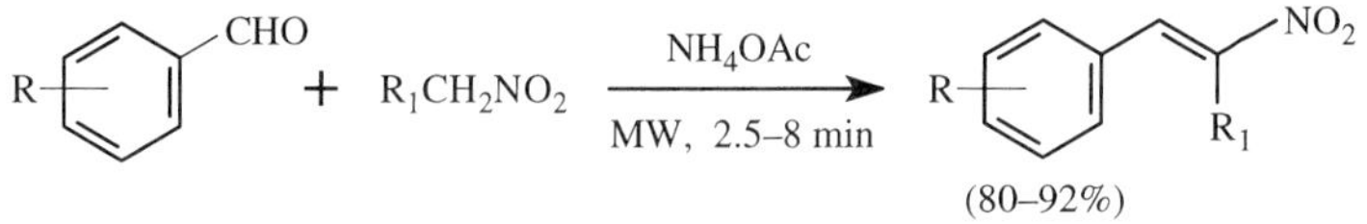

R-⌬-CHO + R₁CH₂NO₂ →[NH₄OAc, MW, 2.5–8 min] R-⌬-CH=C(NO₂)R₁ (80–92%)

R = H, 4-OH, 3,4-(OMe)₂, 3-OMe-4-OH, 1-naphthyl, 2-naphthyl : R₁ = H
R = H, 4-OH, 4-OMe, 3,4-(OMe)₂, 3-OMe-4-OH : R₁ = Me

Fig. 12.21 Synthesis of nitroalkenes.

The Henry reaction (condensation of carbonyl compounds with nitroalkanes to afford nitroalkenes) also proceeds rapidly *via* this MW approach and requires only catalytic amounts of ammonium acetate in reactions involving neat reactants, thus avoiding the use of the large excess of polluting nitrohydrocarbons normally employed (Fig. 12.21).[24]

The reduction, oxidation and cyloaddition reactions emanating from α, β-unsaturated nitroalkenes provide easy access to a vast array of functionalities that include nitroalkanes, N-substituted hydroxylamines, amines, ketones, oximes, and α-substituted oximes and ketones.[33] Consequently, there are numerous possibilities of using these *in situ* generated nitroalkenes for the preparation of valuable building blocks and synthetic precursors.

12.3.2 Expedient synthesis of heterocylic compounds

12.3.2.1 Synthesis of isoflavenes

Isofalv-3-enes, which possess a chromene nucleus, are well known oestrogens and several derivatives of these oxygen heterocycles have attracted the attention of medicinal chemists over the years. Despite the availability of several methods for the synthesis of chromene derivatives, there is demand for the development of eco-friendly synthetic methods for these derivatives. We have uncovered a facile and general method for the synthesis of isoflav-3-enes substituted with basic moieties at the 2 position (Fig. 12.22).[25] Promising results have been obtained for a convergent one-pot synthesis of heterocyclic systems such as 2-substituted isoflav-3-enes wherein the generation of the enamine derivatives *in situ* and inducing subsequent reactions with *o*-hydroxyaldehydes in the same pot is the key feature (Fig. 12.22).

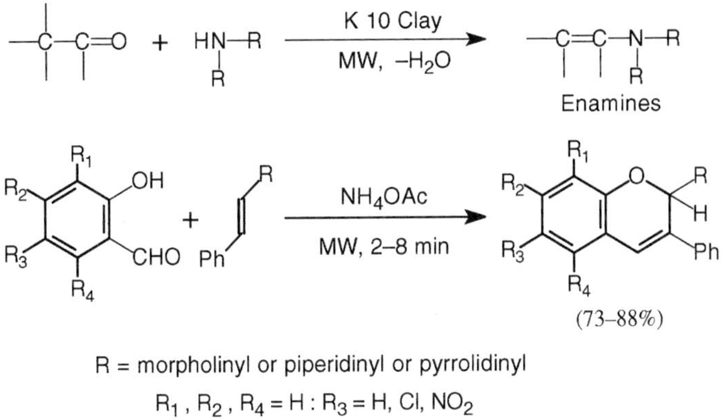

Fig. 12.22 Synthesis of isoflav-3-enes.

The *in situ* generated enamine intermediates have been further elaborated to 2-amino-substituted isoflav-3-enes under solvent-free conditions from readily accessible phenyl acetaldehyde, cyclic amines and salicylaldehydes in presence of ammonium acetate as a catalyst.[25]

12.3.3 Cyclization reactions

12.3.3.1 Synthesis of flavones

Flavonoids are a class of naturally occurring phenolic compounds widely distributed in the plant kingdom, the most abundant being the flavones. Members of this class display a wide variety of biological activities and have been useful in the treatment of various diseases. Flavones have been accessible by various methods such as Allan–Robinson synthesis, synthesis from chalcones and *via* an intramolecular Wittig strategy.[26] The most prevalent method, however, involves the Baker–Venkataraman rearrangement, where *o*-hydroxyacetophenone is benzoylated to form the benzoyl ester followed by the treatment with base (pyridine/KOH) to effect an acyl group migration, forming a 1,3-diketone.[26] The diketone formed is then cyclized under strongly acidic conditions using sulfuric acid and acetic acid to deliver the flavone. Therefore, opportunity exists for the development of an expedient approach using benign and readily available starting materials.

A solvent-free synthesis of flavones is described here which simply involves the microwave irradiation of *o*-hydroxydibenzoylmethanes

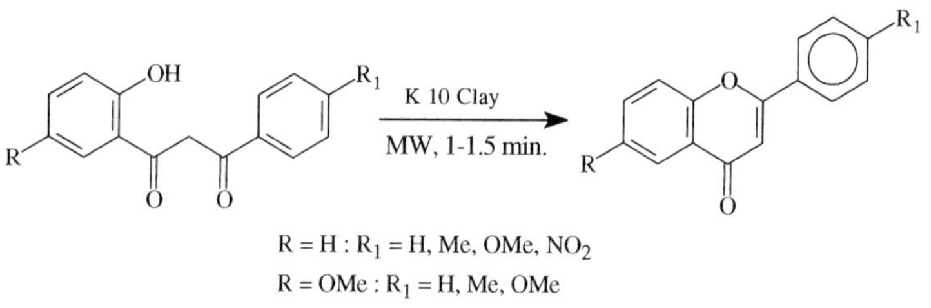

Fig. 12.23 Synthesis of cyclized flavones.

adsorbed on montmorillonite K 10 clay for 1–1.5 min. A rapid and exclusive formation of cyclized flavones occurs in good yields (Fig. 12.23).[26]

12.3.3.2 Synthesis of hydroquinolones

In another solvent-free cyclization reaction using montmorillonite K 10 clay under microwave irradiation conditions, readily available 2′-aminochalcones provide easy access to 2-aryl-1,2,3,4-tetrahydro-4-quinolones,[27] valuable precursors for medicinally important quinolones (Fig. 12.24).

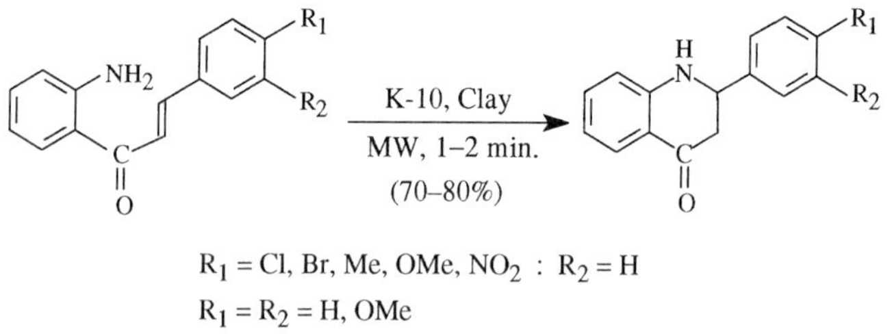

Fig. 12.24 Synthesis of 1,2,3,-4 tetrahydro-4-quinolones.

12.4 Reduction reactions

12.4.1 Borohydride reduction of carbonyl compounds

Relatively inexpensive sodium borohydride ($NaBH_4$) has been extensively used as a reducing agent in view of its compatibility with protic solvents and its safe nature. The solid state reduction of ketones has also

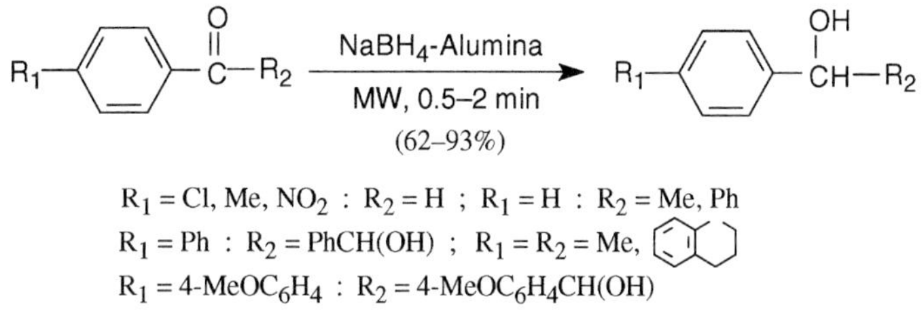

$$R_1\text{-}C_6H_4\text{-}\underset{O}{\overset{\parallel}{C}}\text{-}R_2 \xrightarrow[\text{MW, 0.5-2 min}]{\text{NaBH}_4\text{-Alumina}} R_1\text{-}C_6H_4\text{-}\underset{OH}{\overset{|}{CH}}\text{-}R_2$$

(62–93%)

R_1 = Cl, Me, NO_2 : R_2 = H ; R_1 = H : R_2 = Me, Ph
R_1 = Ph : R_2 = PhCH(OH) ; R_1 = R_2 = Me, (naphthalene)
R_1 = 4-MeOC$_6$H$_4$: R_2 = 4-MeOC$_6$H$_4$CH(OH)

Fig. 12.25 Reduction of carbonyl compounds.

been achieved by mixing with $NaBH_4$ and storing the mixture in a dry box for 5 days. The major disadvantage in the heterogeneous reaction with $NaBH_4$ is that solvent reduces the reaction rate while in the solid state the reaction time is too long (5 days) for it to be of any practical utility.[28]

A facile method for the reduction of aldehydes and ketones has been developed that uses alumina supported $NaBH_4$ and proceeds in the solid state using microwaves.[28] The process in its entirety involves a simple mixing of carbonyl compound with (10%) $NaBH_4$–alumina in solid state and irradiating the mixture in a MW oven for 0.5–2 min (Fig. 12.25).

The useful chemoselective feature of the reaction is apparent from the reduction of *trans*-cinnamaldehyde (cinnamaldehyde/$NaBH_4$-alumina, 1 : 1 mol equiv.); the olefinic moiety remains intact and only the aldehyde functionality is reduced in a facile reaction that occurs at room temperature.

No side product formation is observed in any of the reactions investigated and no reaction takes place in the absence of alumina. Further, the recovered alumina can be recycled by mixing with fresh borohydride and reused for subsequent reductions without any loss in activity. The air used for cooling the magnetron ventilates the microwave cavity thus preventing any ensuing hydrogen from reaching explosive concentrations.

12.4.1.2 Reductive alkylation of amines

Reductive amination of carbonyl compounds has been well documented with sodium cyanoborohydride, sodium triacetoxyborohydride and $NaBH_4$ coupled with sulfuric acid which either produce waste stream or involve the use of corrosive acids. Capitalizing on the knowledge base

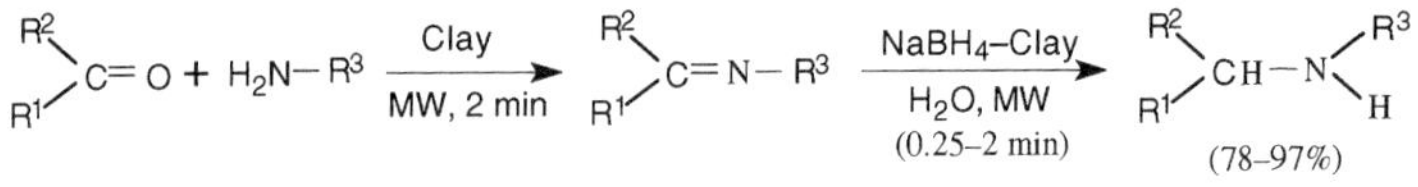

$$R^2\!\!\diagdown\!\!{}_{R^1}\!\!\diagup\!\!C{=}O + H_2N{-}R^3 \xrightarrow[\text{MW, 2 min}]{\text{Clay}} R^2\!\!\diagdown\!\!{}_{R^1}\!\!\diagup\!\!C{=}N{-}R^3 \xrightarrow[\text{H}_2\text{O, MW}]{\text{NaBH}_4\text{–Clay}} R^2\!\!\diagdown\!\!{}_{R^1}\!\!\diagup\!\!CH{-}N\!\diagdown\!\!{}_{H}\!\!\diagup\!\!{}^{R^3}$$
(0.25–2 min)   (78–97%)

R_1 = i-Pr, Ph, 2-HOC$_6$H$_4$, 4-MeOC$_6$H$_4$, 4-NO$_2$C$_6$H$_4$: R_2 = H : R_3 = Ph
R_1 & R_2 = –(CH$_2$)$_5$– : R_3 = Ph ; R_1 & R_2 = –(CH$_2$)$_6$– : R_3 = n-Pr
R_1 = 4-ClC$_6$H$_4$: R_2 = H : R_3 = 2-HOC$_6$H$_4$; R_1 = R_2 = Et : R_3 = Ph
R_1 = n-C$_5$H$_{11}$: R_2 = Me : R_3 = Morpholine, Piperidine ; R_1 = i-Pr : R_2 = H : R_3 = n-C$_{10}$H$_{21}$

Fig. 12.26 Reductive amination of carbonyl compounds.

on environmentally benign methods developed in our laboratory, we have now achieved a solvent-free reductive amination of carbonyl compounds using wet montmorillonite K 10 clay supported sodium borohydride that is facilitated by microwave irradiation (Fig. 12.26).[29]

The practical applications of the reducing potential of NaBH$_4$ on mineral surfaces for the reduction of *in situ* generated Schiff's bases have delivered interesting results.[29] The preliminary studies pertaining to the solid state reductive amination of carbonyl compounds on various inorganic solid supports such as alumina, clay, silica etc. are encouraging with especially interesting results obtained on K 10 clay surface.[29] Clay not only behaves as a Lewis acid but provides water from its interlayers that is responsible for the acceleration of the reducing ability of NaBH$_4$.

12.5 Conclusion

The solvent-free approach using microwave irradiation in conjuction with the use of recyclable mineral solid supports provides interesting possibilities for efficient organic manipulations.[34] The chemo-, regio- and stereoselective synthesis of high value chemicals may encourage the participation of chemical industry to pursue the translation of these laboratory experiments to large scale preparations in view of the solvent and chemical conservation.

Acknowledgement

I am grateful for financial support to the Texas Advanced Research Program (ARP) in chemistry (Grant # 003606-023) and the Texas Research Institute for Environmental Studies (TRIES). I am indebted to contributions from several research associates whose names appear in the references and who have made this work possible, especially Dr Rajender Dahiya for his help in the preparation of this manuscript.

References

1. For recent reviews on microwave-assisted chemical reactions see:
   (a) Abramovich, R.A. (1991). Applications of microwave energy in organic synthesis. *Organic Preperation Proceedings International*, **23**, 683–711. (b) Majetich, G. and Hicks, R. (1995) The use of microwave heating to promote organic reactions. *Journal of Microwave Power Elecromagnetic Energy*, **30**, 27–45. (c) Caddick, S. (1995). Microwave assisted organic reactions. *Tetrahedron*, **51**, 10403–32. (d) Strauss, C.R. and Trainor, R.W. (1995) Invited review. Development in microwave-assisted organic chemistry. *Australian Journal of Chemistry*, **48**, 1665–92. (e) Bose, A.K., Banik, B.K., Lavlinskaia, N., Jayaraman, M. and Manhas, M.S. (1997) MORE chemistry in a microwave. *Chemtech*, **27**, 18–24. (f) Varma, R.S. (1999). Solvent-free organic syntheses using supported reagents and microwave irradiation. *Green Chemistry*, 43–55. (g) Varma, R.S. (1999). Solvent-free organic syntheses on mineral supports using microwave irradiation. *Clean Products and Processes*, **1**, 132–47.
2. Giguere, R.J., Namen, A.M., Lopez, B.O., Arepally, A., Ramos, D.E., Majetich, G. and Defrauw, J. (1987). *Tetrahedron Letters*, **28**, 6553–56.
3. Varma, R.S. (1997). In *Microwaves: theory and application in material processing IV*, American Ceramic Society, Ceramic Transactions, (ed. D.E. Clark, W.H. Sutton and D.A. Lewis), **80**, pp. 357–65.

Cleavage–deprotection reactions

4. Varma, R.S. and Meshram, H.M. (1997). Solid state deoximation with ammonium persulfate-silica gel: Regeneration of carbonyl compounds using microwaves. *Tetrahedron Letters*, **38**, 5427–28.
5. Varma, R.S. and Saini, R.K. (1997). Solid state dethioacetalization using clayfen. *Tetrahedron Letters*, **38**, 2623–4.
6. Varma, R.S., Varma, M. and Chatterjee, A.K. (1993). Microwave-assisted deacetylation on alumina: A simple deprotection method. *Journal of Chemical Society, Perkin Transactions 1*, 999–1000.

7. Varma, R.S., Chatterjee, A.K. and Varma, M. (1993). Alumina-mediated deacetylation of benzaldehyde diacetates. A simple deprotection method. *Tetrahedron Letters*, **34**, 3207–10.
8. Varma, R.S., Lamture, J.B. and Varma, M. (1993). Alumina-mediated cleavage of *t*-butyldimethylsilyl ethers. *Tetrahedron Letters*, **34**, 3029–32.
9. Varma, R.S., Chatterjee, A.K. and Varma, M. (1993). Alumina-mediated microwave thermolysis: A new approach to deprotection of benzyl esters. *Tetrahedron Letters*, **34**, 4603–06.
10. Varma, R.S., Dahiya, R. and Saini, R.K. (1997). Solid state regeneration of ketones from oximes on wet silica supported sodium periodate using microwaves. *Tetrahedron Letters*, **38**, 8819–20.
11. Varma, R.S. and Meshram, H.M. (1997). Solid state cleavage of semicarbazones and phenylhydrazones with ammonium persulfate-clay using microwave or ultrasonic irradiation. *Tetrahedron Letters*, **38**, 7973–6.
12. Varma, R.S. and Kumar, D. (1999). Solventless regeneration of ketones from thioketones using clay supported nitrate salts and microwave irradiation. *Synth. Commun.*, **29**, 1333–40.

Oxidation reactions

13. Varma, R.S. and Dahiya, R. (1997). Microwave-assisted oxidation of alcohols under solvent-free conditions using clayfen. *Tetrahedron Letters*, **38**, 2043-4.
14. Varma, R.S., Saini, R.K. and Dahiya, R. (1997). Active manganese dioxide on silica: Oxidation of alcohols under solvent-free conditions using microwaves, *Tetrahedron Letters*, **38**, 7823–24.
15. Varma, R.S. and Dahiya, R. (1998). Copper(II) nitrate on clay (claycop)-hydrogen peroxide: Selective and solvent-free oxidations using microwaves. *Tetrahedron Letters*, **39**, 1307–08.
16. Varma, R.S., Saini, R.K., (1998). Wet alumina supported chromium(VI) oxide: Selective oxidation of alcohols in solventless system. *Tetrahedron Letters*. **39**, 1481–82.
17. Varma, R.S., Dahiya, R. and Saini, R.K. (1997). Iodobenzene diacetate on alumina: oxidation of alcohols in solventless system using microwaves. *Tetrahedron Letters*, **38**, 7029–32.
18. Varma, R.S., Kumar, D. and Dahiya, R. (1998). Solid state oxidation of benzoins on alumina supported copper(II) sulfate under microwave irradiation. *Journal of Chemical Research (S)*, 324–25.

19. Varma, R.S., Dahiya, R. and Kumar, D. (1998). Solvent-free oxidation of benzoins using Oxone® on wet alumina under microwave irradiation. *Molecules Online*, **2**, 82–85.
20. Varma, R.S., Saini, R.K. and Meshram, H.M. (1997). Selective Oxidation of Sulfides to Sulfoxides and Sulfones by Microwave Thermolysis on Wet Silica-Supported Periodate. *Tetrahedron Letters*, **38**, 6525–8.
21. Varma, R.S., Saini, R.K. and Dahiya, R. (1998). Selective oxidations using alumina supported iodobenzene diacetate under solvent-free conditions. *Journal of Chemical Research (S)*, 120–21.

Condensation–cyclization reactions

22. Varma, R.S., Dahiya, R. and Kumar, S. (1997). Clay catalyzed synthesis of imines and enamines under solvent-free conditions using microwave irradiation. *Tetrahedron Letters*, **38**, 2039–42.
23. Varma, R.S. and Dahiya, R. (1997). Microwave-assisted facile synthesis of imines and enamines using Envirocat EPZG® as a catalyst. *Synlett*, 1245–6.
24. Varma, R.S., Dahiya, R. and Kumar, S. (1997). Microwave-assisted Henry reaction: Solventless synthesis of conjugated nitroalkenes. *Tetrahedron Letters*, **38**, 5131–4.
25. Varma, R.S. and Dahiya, R. (1998). An expeditious and solvent-free synthesis of 2-amino substituted isoflav-3-enes using microwave irradiation. *Journal of Organic Chemistry*, **63**, 8038–41.
26. Varma, R.S., Saini, R.K. and Kumar D. (1998). An expeditious synthesis of flavones on clay using microwaves. *Journal of Chemical Research (S)*, 348–49.
27. Varma, R.S. and Saini, R.K. (1997). Microwave-assisted isomerization of 2'-amino-chalcones on clay: An easy route to 2-aryl-1,2,3,4-tetrahydro-4-quinolones. *Synlett*, 857–58.

Reduction reactions

28. Varma, R.S. and Saini, R.K. (1997). Microwave-assisted reduction of carbonyl compounds in solid state using sodium borohydride supported on alumina. *Tetrahedron Letters*, **38**, 4337–38.
29. Varma, R.S. and Dahiya, R. (1998). Sodium borohydride on wet clay: Solvent-free reductive amination of carbonyl compounds using microwaves. *Tetrahedron*, **54**, 6293–98.

30. (a) Marrero-Terrero, A.L. and Loupy, A. (1996). Synthesis of 2-oxazolines from carboxylic acids and α,α,α-tris (hydroxymethyl)methylamine under microwaves in solvent-free conditions. *Synlett*, 245–46. (b) Villemin, D. and Labiad, B. (1990). Clay catalysis: dry condensation of barbituric acid with aldehydes under microwave irradiation. *Synth. Commun.*, **20**, 3333–37. (c) Villemin, D. and Alloum, A.B. (1990). Potassium fluoride on alumina: condensation of 1,4-diacetylpiperazine-2,5-dione with aldehydes. Dry condensation under microwave irradiation. Synthesis of albonursin and analogues. *Synth. Commun.*, **20**, 3325–31. (d) Villemin, D.; Alloum, A.B. (1991). Dry reaction under microwave: condensation of sulfones with aldehydes on KF-alumina. *Synth. Commun.*, **21**, 63–68. (e) Lerestif, J.M., Bazureau, J.P. and Hamelin, J. (1995). Efficient and practical new synthesis of 2-oxazolines: 1,2-bis- and 1,3-bis (oxazolinyl)benzenes by [3 + 2] cycloaddition using solvent-free conditions. *Synlett*, 647–49.
31. Nelson, D.A., Devin C., Hoffmann, S. and Lau, A. (1997). Division of Chemical Education, *Abstr. No.* 101, ACS National Meeting, San Francisco, April 13–17.
32. Balogh, M. and Laszlo P. (1993). *Organic chemistry using clays*. Springer, Berlin.
33. (a) Varma, R.S. and Kabalka, G.W. (1986). Nitroalkenes in the synthesis of heterocyclic compounds. *Heterocycles*, **24**, 2645–77. (b) Kabalka, G.W. and Varma, R.S. (1987). Synthesis and selected reductions of conjugated nitroalkenes. A review. *Organic Preperation Proceedings International*, **19**, 283–328. (c) Kabalka, G.W., Guindi, L.H.M. and Varma, R.S. (1990). Selected reductions of conjugated nitroalkenes. *Tetrahedron*, **46**, 7443–57.
34. Loupy, A., Petit, A., Hamelin, J., Texier-Boullet, F., Jacquault, P., and Mathe D. (1999). New solvent-free organic synthesis using focused microwaves. *Synthesis*, 1213–34.

13 Organic chemistry via biocatalysis

Andrew Schmid and Bernard Witholt

13.1 Introduction

Biocatalysts have the potential to perform highly specific chemical transformations. Regio- and stereospecific functionalizations of organic molecules, unparalleled by chemical processes, can often be achieved. By conducting reactions in a highly directed and controlled manner, the number of processing steps leading to the synthesis of a desired chemical molecule can be reduced significantly and the formation of unwanted side products is avoided. These effects thus make substantial contributions to reducing the time, the amounts of starting material and solvents as well as the energy required for the synthesis of a given product. We therefore claim that applying biocatalysis for manufacturing valuable chemicals will contribute to environmentally more friendly processes without necessarily increasing operating costs, hence the motivation to develop such bioprocesses.

Organic chemistry is commonly conducted in homogeneous or heterogeneous reaction mixtures containing organic solvents to dissolve hydrophobic reactants and products. Thus, handling of apolar compounds is facilitated and product purification is simplified, due to the volatility of the solvents. Biocatalysts, and especially whole cell biocatalysts, require an aqueous environment, in order to maintain functionality. This apparent contradiction can be resolved by carrying out biocatalytic chemical transformations of hydrophobic compounds in two-liquid phase media. In such whole cell two-liquid phase cultures, cells are grown in an aqueous medium containing water-soluble growth substrates, such as sugars and inorganic salts, while the hydrophobic substrates and/or products are dissolved in a second, apolar, organic solvent phase, typically amounting to 10 to 50% of the total liquid volume.

Such organic solvents can lead to toxic effects for the cells due to hydrophobic molecules partitioning into cell membranes[2,24,25,101,102,110,114] as well as due to physical forces exerted by the organic–aqueous interface.[12,49,81,116] However, micro-organisms respond to the exposure to organic solvents by a number of physiological adaptations to improve their resistance towards apolar compounds, for instance by increasing cell-membrane hydrophobicity and decreasing membrane fluidity.[1,2,8,24,25,84,102,113,114] To further reduce the toxic effects of organic solvents, operational measures can be taken. These include maintaining cells in an optimally, metabolically active state,[39] using optimal concentrations of divalent cations in the aqueous medium,[2,28,48,82,87] immobilizing cells on solid carriers,[13,19,21,32,47,50,59,60] selecting optimal solvents[7,26,29,58,95,107,118,119] and selecting highly resistant strains as well as genetically engineering production strains to increase their solvent tolerance.[4–6,69,75,76,88]

So, although organic solvents added to microbial cultures can put a major burden on the cells, means do exist to minimize these toxic effects, enabling an efficient operation in two-liquid phase systems. In this paper we discuss the main incentives and advantages for performing biotransformation reactions in two-liquid phase systems rather than in conventional single-liquid phase, aqueous fermentations.

13.2 Two-liquid phase bioprocesses

13.2.1 Liquid hydrophobic substrates

Numerous microbial liquid suspension cultures have been reported that utilize apolar liquids as growth substrates or as starting material for biotransformations (Table 13.1). Unlike water soluble substrates, such as glucose and mineral salts, many hydrophobic substrates, especially long chain aliphatic compounds, can be added to microbial cultures in excess amounts, without causing inhibitory effects. This is due to their limited solubilities in the aqueous phase, which are independent of the organic

Table 13.1 Bioprocesses where relatively non-toxic, hydrophobic substrates are added as a second liquid phase to microbial cultures.

Organic solvent	Strain	Product of bioprocess	Reference
Medium chain length alkanes	*Pseudomonas oleovorans*	poly-3-hydroxyalkanoates	48,85,86
Long chain fatty acids	*Pseudomonas putida*	poly-3-hydroxyalkanoates	52
Alkanes and alkenes	recombinant *Escherichia coli*	mono- and dicarboxylic acids	38–40, 94, 100
Triglycerides	*Pseudomonas putida*	fatty acids	66
Castor oil	*Tyromyces sambuceus*	γ-decalactone	56
Isoprenoid hydrocarbons	*Rhodococcus sp.*	oxidized derivatives	77
Medium and long chain alkanes	various bacteria and yeast	biosurfactants	31, 34, 46, 64, 104, 116
Paraffin oils	various yeasts and bacteria	single cell protein (SCP)	35, 67, 99
Fossil fuels	*Rhodococcus sp.*	desulfurized fossil fuels	62, 105
Oil-contaminated water	mixed cultures	purified water	73

to aqueous liquid–liquid volume ratio. Therefore, apolar substrates can be added in bulk amounts avoiding the need for a continuous substrate feed, thus simplifying reactor operation. In some systems, the addition of an apolar liquid substrate in excessive quantities can be a necessity to achieve high process efficiencies.[70] By increasing the volume fraction of the dispersed phase (usually the organic phase) the liquid–liquid interfacial area is increased, and the maximal attainable mass transfer rate from the apolar liquid to the cells is enhanced. For example, the hydrolysis rate of water-immiscible menthyl acetate by *Bacillus subtilis* could be increased two-fold by increasing the organic phase volume fraction from 10% to 50% (v/v).[115]

13.2.2 'Single organic-liquid phase' cultures

An extreme application of organic solvents in whole cell cultures exists where wet cells, with water contents in the range of 0.8 ml per gram of dry cell mass, are suspended in a continuous apolar liquid.[10] A major benefit of such systems, analogous to that of systems using isolated enzymes,[14,45] is the possibility to shift the chemical equilibrium of a biotransformation reaction. For example, the product yield of 1-nitro-2-phenyl-ethane via dehydrogenation of nitrostyrene, could be improved by cultivating wet baker's yeast cells suspended in a continuous liquid phase of light petroleum.[10] However, it is not clear whether metabolic activity, needed in oxido–reduction processes for cofactor regeneration, can be maintained over a prolonged period of time. Cofactors are electron acceptors/donors in biological redox reactions and are recycled intracellularly by virtue of cellular metabolism. The continuous supply of polar substrates and pH control required for optimal cell activity may be severely limited due to the cells being isolated in an aqueous microenvironment. A promising way to overcome this limitation, at least in part, has been indicated by Pinheiro, who showed that dehydrogenation reaction rates could be enhanced by adding solvent soluble quinones as external electron acceptors.[83]

13.2.3 Gaseous hydrophobic substrates

Two-liquid phase cultures have also been used to convert or degrade gaseous substrates dissolved in apolar carrier solvents. For instance, propene is oxidized by *Mycobacterium* sp. to propane oxide in a hydrocarbon two-liquid phase medium.[17,18] A similar approach has been used for the biological treatment of waste gases, where poorly water soluble pollutants adsorbed by an organic solvent in a spray tower are degraded in a continuous two-liquid phase microbial culture.[22]

13.2.4 Solid hydrophobic substrates and products

Numerous bioconversions and biodegradations of solid, hydrophobic compounds have been successfully performed in two-liquid phase systems, many of which involve steroid conversions.[33,36,41,47,90] In many cases higher reaction rates and product yields were attained in two-liquid phase cultures, compared to cultures where the solid substrate was suspended in the aqueous medium (Table 13.2). A number of effects are responsible for these improvements.

Coprecipitation of the hydrophobic substrates and products as well as precipitations on solid cell carrier material are avoided by keeping the compounds in a dissolved state. Thus, substrate yields and productivities can be improved.[32,97,103] In addition, complex substrate feed strategies,

Table 13.2 Biotransformations and biodegradations of solid, hydrophobic compounds in two-liquid phase cultures.

Strain	Process	Organic solvent volume fraction	Activity (2LP:1LP)[1]	Reference
Pseudomonas sp.	cholesterol oxidation	10% p-xylene/diphenylmethane	7	3
Arthrobacter simplex	cholesterol oxidation	25% CCl$_4$	4.3	71
Nocardia sp.	cholesterol oxidation	50% CCl$_4$	35 g l^{-1} h^{-1} : 0	20
Pseudomonas aeruginosa	biodegradation of phenanthrene	20% heptamethylnonane	1.7–2.9	63
mixed culture of Pseudomonas sp. Arthrobacter sp. Alcaligenes sp.	biodegradation of trichlorophenol	20% silicon oil	2	8

1: Volumetric production/biodegradation rate in two-liquid phase culture versus solid–aqueous suspension culture system.

otherwise necessary to avoid substantial losses through precipitation, can be avoided, due to the high substrate solubilities in the organic solvent.

It appears that faster apolar substrate transfer rates are attainable in two-liquid phase cultures containing a dispersed organic liquid, compared to solid suspension cultures. This effect is mainly due to a larger interfacial area between the apolar phase and the aqueous medium in liquid–liquid systems. For instance, higher degradation rates of phenanthrene were obtained in two-liquid phase microbial cultures compared to cultures where equal amounts of the solid substrate were suspended in the aqueous medium.[63]

Increased cell hydrophobicity, a cellular reaction to the presence of hydrophobic organic solvents, has been observed to improve the interaction of cells with the apolar compounds dissolved in the organic phase. As a result, a higher bioavailability and a more efficient utilization of hydrophobic substrates have been observed.[7,8]

13.3 Toxicity

Two-liquid phase processes have been used to continuously extract inhibitory products or to protect cultures from toxic, apolar substrates. Inhibitory effects, otherwise observed in aqueous, single-liquid phase cultivations are reduced by the partitioning of toxic, apolar compounds into the organic liquid phase. Due to the reduced concentrations in the biocatalyst-containing aqueous phase, cells perform better and high conversion rates are achieved despite high titers of toxic chemicals within the two-liquid phase medium. Table 13.3 lists critical concentrations of some inhibitory compounds which have been used in two-liquid phase cultivations. The lowest inhibitory concentrations are in the range of a few mM in aqueous media for aromatic compounds. The lowest critical concentrations in the organic phase lie around 50 to 100 mM. Where the former values depend on the resistance of the microorganisms towards the toxic chemical, the critical concentrations in the organic phase depend on both microbial tolerance and on the partitioning behavior of the toxic chemical between the organic and aqueous phase.[58] As an illustration, Table 13.4 shows organic: aqueous distribution coefficients of various aromatic compounds between octanol and water. The large differences in solute partitioning illustrate the importance of choosing the optimal solvent, since a good partitioning

Table 13.3 Inhibitory concentrations of apolar compounds in single-liquid phase, aqueous cultures and in two-liquid phase, aqueous–organic media.

Chemical	Micro-organism	log P[1]	Crit. aq. conc.[2] [mM]	Crit. org. conc.[3] [mM]	Organic solvent[4]	Reference
Aromatic compounds						
benzene	*Pseudomonas sp*	2.0	1–3			108
toluene	*Escherichia coli*	2.5	1–4	380–470	n-tetradecane	11, 107
	Pseudomonas putida		2.4	< 900	n-tetradecane	26, 117
styrene	*Escherichia coli*	3.0	2.5	180–350	n-octane, n-dodecane	118, 119
	Nocardia corallina			260–350	n-hexadecane	42
styrene-epoxide	*Nocardia corallina*	1.2		80	n-hexadecane	42
benzyl-alcohol	*Pichia pastoris*	1.0	280			58
Aliphatic compounds						
ethanol	*Saccharomyces cerevisiae*	−0.24	400–800			74, 93, 98
butanol	*Clostridium acetobutylicum*	0.8	70–140	270–400	oleyl-alcohol	54, 92
octanol	*Pseudomonas oleovorans*	2.9		95	n-octane	23

Table 13.3 (continued)

Chemical	Micro-organism	log P[1]	Crit. aq. conc.[2] [mM]	Crit. org. conc.[3] [mM]	Organic solvent[4]	Reference
1-octene	Nocardia corallina	4.3		2300	n-hexadecane	57
1,2-epoxy-octane	Nocardia corallina	2.6		60	n-hexadecane	42
	Pseudomonas oleovorans			> 1300	1-octene	29
7,8-epoxy-1-octene	Pseudomonas oleovorans	2.4	140	> 500	1,7-octadiene	96

1: obtained from [65,89].
2: critical aqueous concentration: concentration in aqueous phase where toxic effects were observed.
3: critical organic concentration: concentration in the organic solvent of a two-liquid phase cutlure, where toxic effects were observed.
4: water immiscible solvent used in two-liquid phase cultivation.

Table 13.4 Distribution coefficients of different benzene derivatives between octanol water at 20 to 40 °C.[65,89]

Solute	Distribution coefficient [M M^{-1}]
benzene	100
phenol	32
dihydroxybenzene	10
toluene	320
methoxybenzene	130
benzoic acid	60
benzaldehyde	32
benzylalcohol	10
ethylbenzene	1300
styrene	1000
ethoxybenzene	50
2-phenyl-acetic acid	25
2-phenyl ethanol	25
styrene epoxide	16
2-phenyl-ethanal	10
biphenyl	10'000
hydroxy-biphenyl	3'000
dihydroxy-biphenyl	1'000

into the extractant is desirable. A rule of thumb is that extraction is better the more the solvent physically resembles the solute. Unfortunately the consequence of this fact is that solvents with the best partitioning characteristics are often as toxic to the cells as the product to be extracted, making compromises necessary in choosing extractants. Solvents which show no physical resemblance with the solute, but which specifically interact with the solute, via hydrogen bonding or Lewis acid–base interactions have been used for the extraction of various metabolites (e.g. acetic acid, citric acid or penicillin). However, these alkylated amines, phosphins or chelators can be toxic, and usually a strong decrease of the organic: aqueous partitioning coefficient at solute concentrations above the mM range is observed.

13.3.1 Toxic substrates

Substrate inhibition could be eliminated by diluting toxic starting material in apolar carrier solvents in bioconversions of liquid aromatic compounds,[42,58,109,118,119] olefins[29,57] and unsaturated alcohols.[78–80] An additional advantage of two-liquid phase operation in many of these systems is that inhibitory, hydrophobic products formed are also extracted into the organic solvent (see below). However, it remains to be seen if two-liquid phase operation will also be valuable in optimized production processes, for conversions of toxic substrates into non-toxic or hydrophilic products, such as the biotransformation of toluene to toluene-cis-glycol.[26,107] Controlled substrate addition at rates that avoid substrate concentrations from reaching inhibitory levels are technically and operationally feasible in conventional single-liquid phase fed-batch cultures in laboratory as well as large scale reactors.[11,61] This is achieved either by supplying volatile compounds via the gas phase[51,55,108,111] or via controlled liquid feeds.[9,16,43] Thus, it is likely that many systems cultivated in two-liquid phase media during initial process development will be operated without a distinct organic phase once optimized and scaled up. Nevertheless, in processes with continuously changing conditions, as prevalent in batch and fed-batch cultures and where concentrations must in addition be controlled within a narrow range, a two-liquid phase operation may prove to be indispensable in buffering fluctuations of substrate uptake rates. A system where this might apply could be the biosynthesis of functionalized poly-3-hydroxyalkanoates from toxic precursors (Jung, K., in preparation).

13.3.2 Toxic products: *in-situ* product extraction

The continuous extraction of inhibitory products by a second liquid phase presents a promising tool to attain higher volumetric productivities in biosynthesis and biotransformation processes. By extending the cultivation time before inhibiting product concentrations are reached, higher product titers are attained, and volumes of liquid streams are reduced, due to the higher substrate concentrations and cell densities reached. These two factors enable a more efficient use of the whole cell biocatalyst and reduce the energy requirements of processing, leading to a reduction of production costs.[27,91] Compounds successfully produced in such *in situ* extraction processes include aliphatic and aromatic epoxides,[29,30,42,57,58,118,119] the alcohols ethanol,[74,93] butanol[44,54,92,93,112] and octanol,[15] optically pure derivatives of citronellol[78,79] and organic acids.[68,120]

13.4 Advantage of *in-situ* extraction of biotransformation products

In general, *in situ* product extraction can be expected to improve significantly process performance if the product is inhibitory at low concentrations, in the mM range, and if it partitions readily into a biocompatible, water immiscible solvent, which can be used as extractant. Figure 13.1 shows the estimated maximum productivity of a two-liquid phase process versus that of a conventional single-liquid phase process as a function of the critical product concentration in the aqueous phase and the product partitioning between the organic and aqueous phase (see appendix for model calculations). The graph indicates that *in situ* product extraction can increase the overall productivities by a factor of two or more in batch culture systems where the product is inhibitory at low concentrations in the aqueous

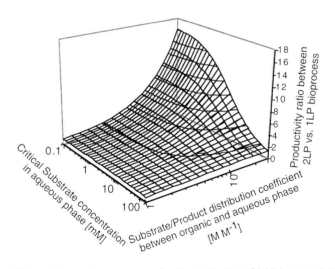

Fig. 13.1 Estimated improvements for the production of inhibitory products by two-liquid phase (2LP) operation compared to production rates attained in conventional single-liquid phase cultivations (1LP). The productivity ratio between 2LP versus 1LP operation is plotted as a function of the substrate distribution coefficient between the organic and aqueous phase, and the critical substrate concentration in the aqueous medium, above which total inhibition is assumed. The ratios were calculated based on Eq. 4, assuming an organic phase volume fraction of 50% (v/v) and constant reactor volumes in 2LP and 1LP processes.

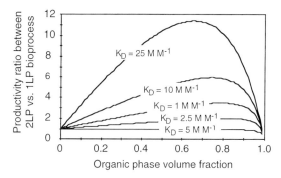

Fig. 13.2 Estimated improvements for the production of inhibitory products by two-liquid phase (2LP) batch operation compared to production rates attained in conventional single-liquid phase batch cultivations (1LP). The productivity ratio between 2LP versus 1LP operation is plotted as a function of the organic phase volume fraction. The ratios were calculated based on Eq. 4, assuming constant reactor volume, substrate-organic: aqueous distribution coefficients between 5 and 25 M M^{-1} as indicated, and assuming a critical substrate concentration in the aqueous medium of 2 mM, above which total inhibition occurs.

phase, in the range of 1 to 10 mM, and where the product partitions well into the extractant, with a distribution coefficient of at least 3. For a compound with an inhibitory concentration of 2 mmol l_{aq}^{-1} in the aqueous phase, produced in a batch process, the optimal organic phase volume fraction is in the range of 0.5 to 0.8, as shown in Fig. 13.2. Significantly improved volumetric production rates over a very wide range of organic to aqueous phase ratios are obtained when the product distribution between the extractant and the aqueous phase is in the order of 20 (mol/mol) or higher. While this applies for most mcl-alkanes and alkenes and their hydroxylated derivatives, other biotransformation products, such as oxidized aromatic compounds, can have distribution coefficients significantly lower than this value, in which case other means for *in situ* product removal must be applied, for instance solid phase extraction (Held *et al.*, submitted).

13.4.1 Potential new two-liquid phase bioprocesses

A large number of biotransformations and biosyntheses catalyzed by whole cells and isolated enzymes are known, and new reactions are continuously being discovered. They include syntheses of a wide range of chemicals such as steroids,[53] various aromatic and aliphatic organic compounds[37] and polymers,[106] which have the potential to serve as chemical precursors for the efficient synthesis of optically pure agents in chemical industries. Many of the substrates and products involved in these processes are hydrophobic and not seldom also toxic to microorganisms. Thus, two-liquid phase cultivation techniques is a potentially valuable tool for maximizing biosynthesis and production rates.

Appendix: comparison of volumetric productivity with and without *in situ* product extraction

Model process

To illustrate and roughly estimate the benefits of *in situ* product extraction, the volumetric reactor productivity over a total production cycle of a batch or fed-batch process was calculated in a simplified model. According to the model a production reactor is inoculated with cells which are grown for 10 h during growth stage up to the initial operating cell density of approximately 10 g l^{-1}. The culture is then induced and substrate for the bioconversion is added. During the following bioconversion stage, cells catalyze the desired biotransformation with an average specific productivity of 17 U $g^{-1}_{\text{cell dry weight}}$. At an average cell density of 10 g l^{-1}, an average volumetric productivity of 10 mmol per liter aqueous medium per hour is maintained during this stage. It was assumed that product extracted into the organic solvent does not affect the cells. The cultivation is terminated when a critical level of product has accumulated within the aqueous phase, at which point inhibitory effects become excessive. Thereupon, during the turnaround stage, the reactor is then emptied and prepared for the next batch, a procedure assumed to take 10 h.[72]

Model calculations

Based on this model, the total amount of product accumulated in a single batch, P [mol], is for low $p^{\text{crit}}_{\text{aq}}$

$$P \leq p^{\text{crit}}_{\text{aq}} \cdot V_r \cdot (1 + (K_D - 1)\Phi). \tag{1}$$

$p^{\text{crit}}_{\text{aq}}$ [mmol l^{-1}_{aq}] is the critical product concentration in the aqueous phase, above which product inhibition becomes excessive and at which point the cultivation is terminated. V_r [m^3] is the reactor working volume, K_D [mol mol^{-1}] the organic: aqueous distribution coefficient of the inhibitory product and Φ the organic phase volume fraction in the reactor.

The average volumetric productivity in terms of product formed per unit of time and reactor volume during the bioconversion stage, $\bar{r}_p^r$ [mmol l^{-1} h^{-1}], is expressed as

$$\bar{r}_p^r = \frac{P}{t_{bc} \cdot V_r} = (1 - \Phi) \cdot \bar{r}_p^{aq} \qquad (2)$$

t_{bc} [h] is the duration of the bioconversion stage and $\bar{r}_p^{aq}$ is the average volumetric productivity of the aqueous phase during this stage, assumed to be 10 mmol l_{aq}^{-1} h^{-1} as stated above.

Eq. 2 indicates that the volumetric reactor productivity during the bioconversion stage is reduced as soon as a second liquid phase is introduced, i.e. $\Phi > 0$. This effect is due to the fact that the organic phase is a catalytically inactive volume, which reduces the volume of the active aqueous phase within the reactor. However, because the second liquid phase continuously removes the inhibitory product from the aqueous phase, the duration of the bioconversion stage is extended before inhibitory concentrations are obtained, leading to a larger final concentration of product. The duration of the bioconversion stage is calculated from Eq. 3, which is derived directly from Eq. 1 and 2.

$$t_{bc} = \frac{P}{V_r \cdot \bar{r}_p^r} \leq \frac{p_{aq}^{crit} \cdot (1 + (K_D - 1)\Phi)}{(1 - \Phi) \cdot \bar{r}_p^{aq}} \qquad (3)$$

The extent of the benefits achieved by continuous *in situ* product removal becomes obvious, when the actual volumetric productivity is considered, i.e. the amount of product formed during a single batch over the whole duration of the bioconversion step, t_{tot}, including the time necessary to grow the cells to the required cell density, t_g, the bioconversion stage itself, t_{bc}, and the time necessary for harvesting and for preparing the following cultivation, during the turnaround stage, t_{to} (Eq. 4).

$$\bar{r}_{p,act}^r = \frac{P}{t_{tot}} = \frac{P}{t_g + t_{bc} + t_{to}} \leq \frac{p_{aq}^{crit} \cdot V_r \cdot (1 + (K_D - 1)\Phi)}{\frac{p_{aq}^{crit} \cdot (1 + (K_D - 1)\Phi)}{(1 - \Phi) \cdot \bar{r}_p^{aq}} + t_g + t_{to}} \qquad (4)$$

References

1. Alexandre, H., Mathieu, B., and Charpentier, C. *Microbiology-Uk* **1996**, *142*, 469–475.
2. Antunes-Madeira, M.C. and Madeira, V.M.C. *Biochim. Biophys. Acta* **1987**, *901*, 61–66.
3. Aono, R., Doukyu, N., Kobayashi, H., Nakajima, H., and Horikoshi, K. *Appl. Environ. Microbiol.* **1994**, *60*, 2518–2523.
4. Aono, R., Negishi, T., Kazuhiko, A., Inoue, A., and Horikoshi, K. *Biosci. Biotech. Biochem.* **1994**, *58*, 1231–1235.
5. Aono, R., Negishi, T., and Nakajima, H. *Appl. Environ. Microbiol.* **1994**, *60*, 4624–4626.
6. Asako, H., Nakajima, H., Kobayashi, K., Kobayashi, M., and Aono, R. *Appl. Environ. Microbiol.* **1997**, *63*, 1428–1433.
7. Ascon-Cabrera, M. and Lebeault, J.M. *Appl. Environ. Microbiol.* **1993**, *59*, 1717–1724.
8. Ascon-Cabrera, M.A. and Lebeault, J.M. *J. Ferm. Bioeng.* **1995**, *80*, 270–275.
9. Badot, R., Coulom, T., Delongeaux, N., Badard, M., and Sibony, J. *Water Sci. Technol.* **1994**, *29*, 329–338.
10. Bak, R.R., McAnda, A.F., Smallridge, A.J., and Trewhella, M.A. *Aust. J. Chem.* **1996**, *49*, 1257–1260.
11. Ballard, D.G.H., Blacker, A.J., Woodley, J.M., and Taylor, S.C. In: *D.P. Mobley, Plastics from microbes: microbial synthesis of polymers and polymer precursors*, **1994**, Hanser Publishers, Munich, 139–168.
12. Bar, R. In: *Conference Proceedings, Biocatalysis in organic media*, **1987**, Wageningen, The Netherlands, Elsevier, 147–153.
13. Bar, R. and Gainer, J.L. *Biotechnol. Prog.* **1987**, *3*, 109–114.
14. Bell, G., Halling, P.J., Moore, B.D., Partridge, J., and Rees, D.G. *Tibtech* **1995**, *13*, 468–473.
15. Bosetti, A., van Beilen, J.B., Preusting, H., Lageveen, R.G., and Witholt, B. *Enzyme Microb. Technol.* **1992**, *14*, 702–708.
16. Brazier, A.J. and Lilly, M.D. *Enzyme Microb. Technol.* **1990**, *12*, 90–94.
17. Brink, L.E.S. and Tramper, J. In: *Conference Proceedings, Biocatalysis in organic media*, 1987, Wageningen, The Netherlands, Elsevier, 133–146.
18. Brink, L.E.S. and Tramper, J. *Enzyme Microb. Technol.* **1987**, *9*, 612–618.
19. Brink, L.E.S., Tramper, J., Luyben, K.C.A.M., and Van't Riet, K. *Enzyme Microb. Technol.* 1988, *10*, 736–743.

20. Buckland, B.C., Dunnill, P., and Lilly, M.D. *Biotechnol. Bioeng.* **1975**, *17*, 815–826.
21. Buitelaar, R.M., Vermue, M.H., Schlatmann, J.E., and Tramper, J. *Biotechnol. Tech.* **1990**, *4*, 415–418.
22. Cesario, M.T., Beeftink, H.H., and Tramper, J. In: *A.J. Dragt and J. von Ham, Biotechniques for air pollution abatement and odour control policies*, **1992**, Elsevier Science Publishers, Amsterdam, 135–142.
23. Chen, Q. *Growth of Pseudomonas oleovorans in two liquid phase media: effects of organic solvents and alk gene expression on the membrane*. Ph.D. thesis, Rijksuniversiteit te Groningen, Netherlands, **1996**.
24. Chen, Q., Janssen, D.B., and Witholt, B. *J. Bacteriol.* **1995**, *177*, 6894–6901.
25. Chen, Q., Nijenhuis, A., Preusting, H., Dolfing, J., Janssen, D.B., and Witholt, B. *Enzyme Microb. Technol.* **1995**, *17*, 647–652.
26. Collins, A.M., Woodley, J.M., and Liddell, J.M. *J. Ind. Microbiol.* **1995**, *14*, 382–388.
27. Daugulis, A.J., Axford, D.B., and McLellan, P.J. *Can. J. Chem. Eng.* **1991**, *69*, 488–497.
28. de Smet, M.J., Kingma, J., and Witholt, B. *Biochim. Biophys. Acta* **1978**, *506*, 64–80.
29. de Smet, M.J., Kingma, J., Wynberg, H., and Witholt, B. *Enzyme Microb. Technol.* **1983**, *5*, 352–360.
30. de Smet, M.J., Wynberg, H., and Witholt, B. *Appl. Environ. Microbiol.* **1981**, *Nov.*, 811–816.
31. Desai, J.D. and Banat, I.M. *Microbiol. Mol. Biol. Rev.* **1997**, *61*, 47–64.
32. Dias, A.C.P., Cabral, J.M.S., and Pinheiro, H.M. *Enzyme Microb. Technol.* **1994**, *16*, 708–714.
33. Doukyu, N., Kobayashi, H., Nakajima, H., and Aono, R. *Biosci. Biotech. Biochem.* **1996**, *60*, 1612–1616.
34. Duvnjak, Z., Cooper, D.G., and Kosaric, N. *Biotechnol. Bioeng.* **1982**, *24*, 165–175.
35. Einsele, A. In: *H. Dellweg, Biotechnology*, **1983**, *Verlag Chemie, Weinheim*, 43–81.
36. Evans, T.W., Kominek, L.A., Wolf, H.J., Henderson, S.L., and Perry, R.E. *Patent* **1984**, EU 0 127 294.
37. Faber, K. *Biotransformations in Organic Chemistry*, **1997**, Springer, Berlin.
38. Favre Bulle, O., Schouten, T., Kingma, J., and Witholt, B. *Bio/Technol.* **1991***, 9, 367–371.*

39. Favre Bulle, O., Weenink, E., Vos, T., Preusting, H., and Witholt, B. *Biotechnol. Bioeng.* **1993**, *41*, 263–272.
40. Favre Bulle, O. and Witholt, B. *Enzyme Microb. Technol.* **1992**, *14*, 931–937.
41. Fernandes, P., Cabral, J.M.S., and Pinheiro, H.M. *Enzyme Microb. Technol.* **1995**, *17*, 163–167.
42. Furuhashi, K., Shintani, M., and Takagi, M. *Appl. Microbiol. Biotechnol.* **1986**, *23*, 218–223.
43. Gbewonyo, K., Buckland, B.C., and Lilly, M.D. *Biotechnol. Bioeng.* **1991**, *37*, 1101–1107.
44. Groot, W.J., Soedjak, H.S., Donck, P.B., van der Lans, R.G.J.M., Luyben, K.C.A.M., and Timmer, J.M.K. *Bioproc. Eng.* **1990**, *5*, 203–216.
45. Gutman, A.L. and Shapira, M. *Adv. in Biochem. Eng. Biotechnol.* **1995**, *52*, 87–128.
46. Haferburg, D., Hommel, R., Claus, R., and Kleber, H. *Adv. Biochem. Eng.* **1986**, *33*, 53–93.
47. Harrop, A.J., Woodley, J.M., and Lilly, M.D. *Enzyme Microb. Technol.* **1992**, *14*, 725–730.
48. Hazenberg, W.M. *Production of poly(3-hydroxyalkanoates) by Pseudomonas oleovorans in two-liquid phase media.* Ph.D. thesis, Eidgenösische Technische Hochschule Zürich, Switzerland, **1997**.
49. Hocknull, M.D. and Lilly, M.D. In: *Conference Proceedings, Biocatalysis in Organic Media*, **1987**, Wageningen, The Netherlands, Elsevier Science Publishers, 393–398.
50. Hocknull, M.D. and Lilly, M.D. *Appl. Microbiol. Biotechnol.* **1990**, *33*, 148–153.
51. Hsieh, J.H. Patent **1989**, US 4 833 078.
52. Huijberts, G.N.M. and Eggink, G. *Appl. Microbiol. Biotechnol.* **1996**, *46*, 233–239.
53. Iizuka, H. and Naito, A. *Microbial conversion of steroids*, **1981** University of Tokyo Press, Springer Verlag, Berlin.
54. Ishii, S., Taya, M., and Kobayashi, T. *Chem. Eng. Japan* **1985**, *18*, 125–130.
55. Jenkins, R.O., Stephens, G.M., and Dalton, H.; *Biotechnol. Bioeng.* **1987**, *29*, 873–883.
56. Kapfer, G.F., Berger, R.G., and Drawert, F. *Biotechnol. Lett.* **1989**, *11*, 561–566.

57. Kawakami, K. In: *H. Okada, A. Tanaka, and H.W. Blanch, Enzyme Engineering X*, **1990**, New York Academy of Sciences, New York, 707–711.
58. Kawakami, K. and Nakahara, T. *Biotechnol. Bioeng.* **1994**, *43*, 918–924.
59. Keay, L., Eberhardt, J.J., Allen, B.R., Scott, C.D., and Davison, B.H. In: *Conference Proceedings, Physiology of immobilized cells*, **1990**, Wageningen, The Netherlands, Elsevier Science Publishers, 539–543.
60. Keweloh, H., Heipieper, H.J., and Rehm, H.G. In: *Conference Proceedings, Physiology of immobilized cells*, **1990**, Wageningen, The Netherlands, Elsevier Science Publishers, 545–550.
61. Kiener, A. Patent **1993**, US 5 236 832.
62. Knapp, R.D., Xu, G.W., Squires, C.H., Monticello, D.G., and Pienkos, P.T. *Abstr. Gen. Meet. Am. Soc. Microbiol.* **1995**, *95*, 374.
63. Köhler, A., Schüttoff, M., Bryniok, D., and Knackmuss, H.J. *Biodegradation* **1994**, *5*, 93–103.
64. Kosaric, N., Gray, N.C.C., and Cairns, W.L. In: *H. Dellweg, Biotechnology: a comprehensive treatise in 8 volumes*, 1983, Verlag Chemie, Weinheim, 576–592.
65. Laane, C., Boeren, S., Vos, K., and Vergeer, C. *Biotechnol. Bioeng.* **1987**, *30*, 81–87.
66. Lee, S.Y. and Rhee, J.S. *Biotechnol. Bioeng.* **1994**, *44*, 437–443.
67. Levi, J.D., Shennan, J.L., and Ebbon, G.L. In: *A.H. Rose, Economic Microbiology*, **1979**, Acad. Press, New York, 361–419.
68. Levy, P.F., Sanderson, J.E., and Wise, D.L. *Biotechnol. Bioeng. Symp.* **1981**, *11*, 239–248.
69. Li, L., Komatsu, T., Inoue, A., and Horikoshi, K. *Biosci. Biotechnol. Biochem.* **1995**, *59*, 2358–2359.
70. Lilly, M.D., Dervakos, G.A., and Woodley, J.M. In: *L.G. Copping, et al., Opportunities in Biotransformations*, **1990**, Elsevier, London, 5–16.
71. Liu, W.H., Horng, W.C., and Tsai, M.S. *Enzyme Microb. Technol.* **1996**, *18*, 184–189.
72. Mathys, R.G. *Bioconversion in two-liquid phase systems: downstream processing. Ph.D. thesis, Eidgenössische Technische Hochschule Zürich, Switzerland,* **1997**.
73. Mattei, G., Rambeloarisoa, E., Giusti, G., Routani, J.F., and Bertrand, J.D. *Appl. Microbiol. Biotechnol.* **1986**, *23*, 302–304.
74. Moritz, J.W. and Duff, S.J.B. *Biotechnol. Bioeng.* **1996**, *49*, 504–511.

75. Moriya, K., Yanagitani, S., Usami, R., and Horikoshi, K. *J. Mar. Biotechnol.* **1995**, *2*, 131–133.
76. Nakajima, H., Kobayashi, H., Negishi, T., and Aono, R. *Biosci. Biotech. Biochem* **1995**, *59*, 1323–1325.
77. Nakajima, K. In: *G.W. Moody and P.B. Baker, Bioreactors and biotransformations*, **1987**, Elsevier Applied Science Publications, London, 219–230.
78. Oda, S., Inada, Y., Kato, A., Matsudomi, N., and Ohta, H. *J. Ferment. Bioeng.* **1995**, *80*, 559–564.
79. Oda, S., Inada, Y., Kobayashi, A., Kato, A., Matsudomi, N., and Ohta, H. *Appl. Environ. Microbiol.* **1996**, *62*, 2216–2220.
80. Oda, S. and Ohta, H. *Biosci. Biotech. Biochem..* **1992**, *56*, 2041–2045.
81. Osborne, S.J., Leaver, J., Turner, M.K., and Dunnill, P. *Enzyme Microb. Technol.* **1990**, *12*, 281–291.
82. Pinheiro, H.M. and Cabral, J.M.S. *Biotechnol. Bioeng.* **1991**, *37*, 97–102.
83. Pinheiro, H.M., Cabral, J.M.S., and Adlercreutz, P. *Biocatalysis* **1993**, *7*, 83–96.
84. Pinkart, H.C., Wolfram, J.W., Rogers, R., and White, D.C. *Appl. Environ. Microbiol.* **1996**, *62*, 1129–1132.
85. Preusting, H., Hazenberg, W., and Witholt, B. *Enzyme Microb. Technol.* **1993**, *15*, 311–316.
86. Preusting, H., van Houten, R., Hoefs, A., van Langenberghe, E.K., Favre Bulle, O., and Witholt, B. *Biotechnol. Bioeng.* **1993**, *41*, 550–556.
87. Ragot, F. and Reuss, M. *Biochem. Eng. Stuttgart* **1991**, 184–187.
88. Ramos, J.L., Duque, E., Huertas, M.J., and Haidour, A. *J. Bacteriol.* **1995**, *177*, 3911–3916.
89. Rekker, R.F. and de Kort, H.M. *Eur. J. Med. Chem.—Chimica Therapeutica* **1979**, *14*, 479–488.
90. Ribeiro, M.H., Cabral, J.M.S., and Fonseca, M.M.R. In: *Conference Proceedings, Biocatalysis in organic media*, **1987**, Wageningen, The Netherlands, Elsevier, 323–329.
91. Roffler, S., Blanch, H.W., and Wilke, C.R. *Biotechnol. Prog.* **1987**, *3*, 131–140.
92. Roffler, S.R., Blanch, H.W., and Wilke, C.R. *Bioproc. Eng.* **1987**, *2*, 181–190.
93. Roffler, S.R., Randolph, T.W., Miller, D.A., Blanch, H.W., and Prausnitz, J.M. In: *B. Mattiason and O. Holst, Extractive Bioconversions*, **1990**, M. Dekker, New York, 133–172.

94. Schindler, J., Viehweg, H., Schmid, R., Weiss, A., Ott, K.H., and Eierdanz, H. *Patent* **1987**, DE 3 540 834 A1.
95. Schmid, A. *Two-liquid phase bioprocess development: interfacial mass transfer rates and explosion safety.* Ph.D. thesis, Eidgenössische Technische Hochschule, Zürich, **1997**.
96. Schwartz, R.D. and McCoy, C.J. *Appl. Environ. Microbiol.* **1976**, *31*, 78–82.
97. Sedlaczek, L., *CRC Crit. Rev. Biotechnol.* **1988**, 7, 187–236.
98. Shabtai, Y., Chaimovitz, S., Freeman, A., Katchalski-Katzir, E., Linder, C., Nemas, M., Perry, M., and Kedem, O. *Biotechnol. Bioeng.* **1991**, *38*, 869–879.
99. Shennan, J.L. In: *R.M. Atlas, Petroleum Microbiology*, 1984, Macmillan, New York, 643–683.
100. Shimizu, S., Jareonkitmongkol, S., Kawashima, H., Akimoto, K., and Yamada, H. *Arch. Microbiol.* **1991**, *156*, 163–166.
101. Sikkema, J., de Bont, J.A.M., and Poolman, B. *J. Biol. Chem.* **1994**, *269*, 8022–8028.
102. Sikkema, J., de Bont, J.A.M., and Poolman, B. *Microbiol. Rev.* **1995**, *59*, 201–222.
103. Silbiger, E. and Freeman, A. *Enzyme Microb. Technol.* **1991**, *13*, 869–872.
104. Singh, M., Saini, V.S., Adhikori, D.K., Desai, J.D., and Sista, V.R. *Biotechnol. Lett.* **1990**, *12*, 743–746.
105. Squires, C.H. *Patent* **1996**, WO 96/17 940.
106. Steinbüchel, A. and Valentin, H.E. *FEMS Microbiol. Lett.* **1995**, *128*, 219–228.
107. Tsai, J.T., Wahbi, L.P., Dervakos, G.A., and Stephens, G.M. *Biotechnol. Lett.* **1996**, *18*, 241–244.
108. van den Tweel, W.J.J., de Bont, J.A.M., Vorage, M.J.A.W., Marsman, E.H., Tramper, J., and Koppejan, J. *Enzyme Microb. Technol.* **1988**, 10, 134–142.
109. van den Tweel, W.J.J., Marsman, E.H., Vorage, M.J.A.W., Tramper, J., and de Bont, J.A.M. In: *G.W. Moody and P.B. Baker, Bioreactors and biotransformations*, **1987**, Elsevier Applied Science Publications, London, 231–240.
110. van der Werf, M.J., Hartmans, S., and van den Tweel, W.J.J. *Appl. Microbiol. Biotechnol.* **1995**, *43*, 590–594.
111. van Ede, C.J. *Bioconversions catalyzed by growing immobilized bacteria.* Ph.D. thesis, Rijksuniversiteit te Groningen, Netherlands, **1994**.

112. Wayman, M. and Parekh, R. *J. Ferm. Technol.* **1987**, *65*, 295–300.
113. Weber, F.J. and de Bont, J.A.M. *Meded. Fac. Landbouwwet. Rijksuniv. Gent* **1994**, *59*, 2295–2392.
114. Weber, F.J. and de Bont, J.A.M. *Biochim. Biophys. Acta* **1996**, *1286*, 225–245.
115. Williams, A.C., Woodley, J.M., Ellis, P.A., Narendranathan, T.J., and Lilly, M.D. *Enzyme Microb. Technol.* **1990**, *12*, 260–265.
116. Witholt, B., de Smet, M.J., Kingma, J., van Beilen, J.B., Kok, M., Lageveen, R.G., and Eggink, G. *Tibtech.* **1990**, *8*, 46–52.
117. Woodley, J.M., Brazier, A.J., and Lilly, M.D. *Biotechnol. Bioeng.* **1991**, *37*, 133–140.
118. Wubbolts, M.G., Favre-Bulle, O., and Witholt, B. *Biotechnol. Bioeng.* **1996**, *52*, 301–308.
119. Wubbolts, M.G., Hoven, J., Melgert, B., and Witholt, B. *Enzyme Microb. Technol.* **1994**, *16*, 887–894.
120. Yabannavar, V.M. and Wang, D.I.C. *Biotechnol. Bioeng.* **1991**, *37*, 1095–1100.

Index

acidic ion-exchange polymers 166
acid-impregnated silica 165
Ag(I)/ZSM-5 zeolite photocatalyst 5
alkane oxidation 128, 145, 148
alkylation reaction 164, 169, 172
alkylation with phenylacetylene 41, 42
ammonium persulfate on silica 226
aromatic polycarbonates 208, 210–11

beta zeolites 169, 177
biocatalysis 245
biodesulfurization 107, 109, 112, 119, 121
biological treatment of waste gases 248
biomimetic catalysis 131
biomimetic properties 139
Biosil method 62–3
biosyntheses 256
biotransformations: 246, 256
   in-situ extraction of products 255
   toxic products 254
   toxic substrates 253
bisphenol A 208
bleaching processes 84–8

Cannizzaro reaction 193, 195–6
carbonic esters 214
catalyst deactivation 177, 192
catalytic oxidation 184
cell immobilization 61, 66
cellulose 84
chlorine dioxide 86–8
chromium trioxide 231
claycop 222, 230

clayfen 222, 226, 228–9
clay-supported iron(III) 221
condensation reactions 234
continuous gas-phase reaction 184
copper sulfate alumina 233
cracking 169
cyclohexane oxygenation 125, 133, 135, 137

D bleaching 86–8
deacylation reactions 223–4
debenzylation reactions 223
decomposition of NO into N_2O 15
decomposition of toxic componds 11
decontamination of polluted water 11
deprotection reactions 222
desulfurization, 4-S pathway 109, 111
desulfurization of diesel oil 113
diastereoselective synthesis of phenylcoumaran dimers 27
dimethylcarbonate: 201
   and fuels 216
   as solvent 215
   production *via*: 202
      carbonylation of methylnitrite 202
      methanol oxycarbonylation 204
   toxicological properties 205
diphenylcarbonate 208

elimination of nitrogen oxides 4
encapsulation of living cells 63
epoxidation 152, 154–5
ethane-1,2-diol oxidation 190

fluorous biphase system 147
fluorous/organic two-phase conditions 145

gas phase oxidation 188.191
gasoline 118, 120, 163
gasoline alkylation 164
glycolic acid 190
gold catalysts 185–8,191

harmless reagents 206
heteropolyacids 166
heteropolyacids catalysts 172–3
hexamethylenediisocyanate via dimethylcarbonate 214
homogeneous versus heterogeneous catalysis 146
horseradish peroxidase 21
 catalyzed oxidative coupling of phenols 32
 catalyzed oxidations 22
hydroboration of alkenes 158
hydroformylation of propene 147
hydrogen peroxide 79, 90–2, 99
hydrogen production 9
hydroxy-acids production 183

industrials ecology 33
inorganic photocatalysts 125–40
iodobenzene diacetate 232
iron porphyrins 126–7, 139–40
isobutane alkylation 165–6
isobutane/2-butane alkylation 163, 167, 170
isocyanaets production via dimethylcarbonate 212

Knoevenagel coumarins synthesis 53
Kraft process 85–6

lactic acid 194

lignin 85, 87, 89, 98
liquid phase oxidation 188
liquid phase photocatalytic reactions 12

metal-template electrophilic substitution 41
methylation reactions with dimethylcarbonate 207
microwave 221
Mn(III)-complexes 151
model calculations in biosyntheses 257
molecular oxygen in green synthesis 125, 128, 184–5
montmorillonite 44, 235
montmorillonitic clays 54

nafion 131, 135, 166–7, 174–5
nafion resin/silica nanocomposites 167
$NaIO_4$ on silica 227
nanocomposite of Nafion resin/silica catalysts 174

oligomerization 169
oxidation of 1,2-diols 183
oxidation of 2,6-dimethylphenol 24
oxidation of 4-methylphenol 22
oxidation of 4-terbutylphenol 23
oxidation reactions 228
oxidative cleavage of C–S bonds 108
oxidative dimerization of E–methyl ferulate 27, 29

palladium on carbon, 5% 185,191,195
perfluorinated solvents 148
perfluoroalkyl-substituted porphyrins 150
perfluorocarbons 149–58

perfluorohexane 156–7
phase-transfer catalysis 146
phenol oxidative coupling 21, 25
photocatalysis for environmental detoxification 3
photocatalysis fundamentals 2
photocatalytic decomposition of NO 16
photo-oxidation processes 127, 130
platinum on carbon, 5% 185,191,195
polycharbonate biphenils 12, 82
poly-chlorinated dibenzofurans 87
poly-chorinated dibenzo-p-dioxins 12, 82, 87
polyoxotungstates 126, 137
porphyrins 149, 150–2
productivity of SiO_2-encapsulated cells 69
propane-1,2-diol oxidation 194
pseudomonas strains 108
pulp and paper industry 79, 84
pulp bleaching 86–7, 94–6

recycling of the catayst 145
reduction of carbon dioxide with water 8
reduction reactions 238
reductive alkylation of amines 239
removal of NOx in the atmosphere 6
removal of offensive odors and disinfection 8

selective monomethylations with dimethylcarbonate 208
semiconducting materials 13
separation 159
sol-gel process 62
solid acid catalysts 163, 175, 177
solid superacid catalysts 168
solventless oxidation reactions 229
solventless reactions 37, 237
substrate specificity 26

sulfated metal oxides 168
synthesis of:
  2H-1-benzopyrans 52
  chromanes 45
  chromenes 49
  coumarins 56
  cyclization reactions 237
  enamines 235
  flavones 237–8
  imines 234
  isoflavenes 236
  N,N'-dialkylureas 40
  N,N'-diphenylureas 38, 39
  ureas 37
  ortho-isopentenylphenols 45

1,2-tungstophosphoric acid 173
2,3,7,8-TCDD 87
taxus diterpenes 61
TCF bleaching 89, 90, 98
thermophilic bacteria 108
titanium oxide photocatalysts: 1, 4, 126, 134–5
  implanted with Cr or V ions 15
  photocatalytic panels in expressway 7
  second-generation photocatalysts 16
  anchored onto zeolites 5
totaly chorinee free processes 79, 88
triflic acid supported 165
trimethylpentanes 164
two-liquid phase bioprocesses 245–6, 255
two-liquid phase cultures 249

undifferentiated cell cultures 64

water splitting 10

zeolite HSZ-360 47–8, 51
zeolites 165, 173
zirconia sulfated catalysts 166, 169, 177

TP 155.2 .E58 G74 2000 c.1

Green chemistry

DATE DUE